Proteomics

Targeted Technology, Innovations and Applications

Edited by

Manuel Fuentes

Department of Medicine
University of Salamanca
Salamanca
Spain

and

Joshua LaBaer

The Biodesign Institute
Arizona State University
Tempe, AZ
USA

Caister Academic Press

Caister Academic Press
Norfolk, UK

www.caister.com

British Library Cataloguing-in-Publication Data
A catalogue record for this book is available from the British Library

ISBN: 978-1-908230-46-1 (hardback)
ISBN: 978-1-908230-62-1 (ebook)

Cover design adapted from images supplied by Maria Gonzalez and Paula Díez.

Contents

Contributors

Juan P. Albar
Proteomics Facility
National Center for Biotechnology
CSIC; and
Carlos III Networked Proteomics Platform
ProteoRed
Madrid
Spain

jpalbar@cnb.csic.es

Fernando J. Corrales
Center for Applied Medical Research (CIMA)
University of Navarra
Pamplona
Spain

fjcorrales@unav.es

Noelia Dasilva
Cancer Research Centre
Department of Medicine
University of Salamanca
Salamanca
Spain

noeliadf@usal.es

Rosa M. Dégano
Proteomics Unit
Cancer Research Centre
Department of Medicine
University of Salamanca
Salamanca
Spain

romade@usal.es

Paula Díez
Cancer Research Centre
Department of Medicine
University of Salamanca
Salamanca
Spain

pauladg@usal.es

Manuel Fuentes
Cancer Research Centre
Department of Medicine
University of Salamanca
Salamanca
Spain

mfuentes@usal.es

Kishore Gollapalli
Wadhwani Research Center for Biosciences and Bioengineering
Department of Biosciences and Bioengineering
Indian Institute of Technology Bombay
Mumbai
India

kishore.aprjc@gmail.com

Rafael Góngora
Cancer Research Centre
Department of Medicine
University of Salamanca
Salamanca
Spain

rgongora@usal.es

Maria Gonzalez-Gonzalez
Cancer Research Centre
Department of Medicine
University of Salamanca
Salamanca
Spain

mariagg@usal.es

Nieves Ibarrola
Proteomics Unit
Cancer Research Centre
Department of Medicine
University of Salamanca
Salamanca
Spain

nibarrola@usal.es

Ricardo Jara-Acevedo
ImmunoStep SL
Cancer Research Centre Building
Salamanca
Spain

rjara@immunostep.com

Joshua LaBaer
The Biodesign Institute
Arizona State University
Tempe, AZ
USA

joshua.labaer@asu.edu

Lars Malmström
Institute of Molecular Systems Biology
Department of Biology
ETH Zurich
Zurich
Switzerland

lars@imsb.biol.ethz.ch

György Marko-Varga
Clinical Protein Science & Imaging
Department of Biomedical Engineering
Lund University
Biomedical Center
Lund
Sweden; and
First Department of Surgery
Tokyo Medical University
Tokyo
Japan

gyorgy.marko-varga@bme.lth.se

J. Alberto Medina-Aunon
Proteomics Facility
National Centre for Biotechnology
CSIC
Carlos III Networked Proteomics Platform
ProteoRed
Madrid
Spain

jamedina@cnb.csic.eu

Johnathan Neiswinger
Department of Pharmacology and Molecular Sciences; and
Center for High-Throughput Biology
Johns Hopkins School of Medicine
Baltimore, MD
USA

jneiswi1@jhmi.edu

Alberto Orfao
Cancer Research Centre
Department of Medicine
University of Salamanca
Salamanca
Spain

orfao@usal.es

Jiang Qian
Department of Ophthalmology
Johns Hopkins School of Medicine
Baltimore, MD
USA

jiang.qian@jhmi.edu

Sandipan Ray
Wadhwani Research Center for Biosciences and Bioengineering
Department of Biosciences and Bioengineering
Indian Institute of Technology Bombay
Mumbai
India

sandipaniitbombay@gmail.com

Sanjeeva Srivastava
Wadhwani Research Center for Biosciences and Bioengineering
Department of Biosciences and Bioengineering
Indian Institute of Technology Bombay
Mumbai
India

sanjeeva@iitb.ac.in

Parvez Syed
Wadhwani Research Center for Biosciences and Bioengineering
Department of Biosciences and Bioengineering
Indian Institute of Technology Bombay
Mumbai
India

parvezsyedsyed@gmail.com

Ákos Végvári
Clinical Protein Science & Imaging
Department of Biomedical Engineering
Lund University Biomedical Center
Lund
Sweden

akos.vegvari@bme.lth.se

Jin Zhang
Department of Pharmacology and Molecular Sciences
Johns Hopkins School of Medicine
Baltimore, MD
USA

jzhang32@jhmi.edu

Heng Zhu
Department of Pharmacology and Molecular Sciences; and
Center for High-Throughput Biology
Johns Hopkins School of Medicine
Baltimore, MD
USA

hzhu4@jhmi.edu

Current books of interest

Antibiotics: Current Innovations and Future Trends	2015
Leishmania: Current Biology and Control	2015
Acanthamoeba: Biology and Pathogenesis (2nd edition)	2015
Microarrays: Current Technology, Innovations and Applications	2014
Metagenomics of the Microbial Nitrogen Cycle	2014
Pathogenic Neisseria: Genomics, Molecular Biology and Disease Intervention	2014
Biofuels: From Microbes to Molecules	2014
Human Pathogenic Fungi: Molecular Biology and Pathogenic Mechanisms	2014
Applied RNAi: From Fundamental Research to Therapeutic Applications	2014
Halophiles: Genetics and Genomes	2014
Molecular Diagnostics: Current Research and Applications	2014
Phage Therapy: Current Research and Applications	2014
Bioinformatics and Data Analysis in Microbiology	2014
The Cell Biology of Cyanobacteria	2014
Pathogenic *Escherichia coli*: Molecular and Cellular Microbiology	2014
Campylobacter Ecology and Evolution	2014
Burkholderia: From Genomes to Function	2014
Myxobacteria: Genomics, Cellular and Molecular Biology	2014
Next-generation Sequencing: Current Technologies and Applications	2014
Omics in Soil Science	2014
Applications of Molecular Microbiological Methods	2014
Mollicutes: Molecular Biology and Pathogenesis	2014
Genome Analysis: Current Procedures and Applications	2014
Bacterial Toxins: Genetics, Cellular Biology and Practical Applications	2013
Bacterial Membranes: Structural and Molecular Biology	2014
Cold-Adapted Microorganisms	2013
Fusarium: Genomics, Molecular and Cellular Biology	2013
Prions: Current Progress in Advanced Research	2013
RNA Editing: Current Research and Future Trends	2013
Real-Time PCR: Advanced Technologies and Applications	2013
Microbial Efflux Pumps: Current Research	2013
Cytomegaloviruses: From Molecular Pathogenesis to Intervention	2013

Full details at www.caister.com

Preface

Targeted proteomics is an exciting emerging approach that uses high-throughput tools to detect proteins of interest with high sensitivity, quantitative accuracy and good reproducibility. The analysis of a pre-defined group of proteins delivers precise and sensitive data to biologists and clinicians, without having to sift through large quantities of data not relevant to the biological question. Recently, important advances in targeted proteomics have accelerated, both in technological methodology and in discoveries. This has provided information regarding specific proteins and their concentrations in a sample, shedding light on the activity of signalling pathways and providing clues about how the changing components fulfil a physiological function.

The adoption of targeted proteomics to study biological and clinical questions is well under way in the biomedical community based on the growing reliability of the measurements made using targeted methods including both mass spectrometry and other proteomics approaches. Although the methods are not yet fully matured, they are improving at a remarkable pace. Key next steps will be to convert them into robust approaches that can transfer easily from one lab to the next. Ideally, each lab studying the same sample will get the same answers.

Currently, researchers in targeted proteomics are working on reducing sample complexity, assessing reproducibility and increasing sensitivity in order to fill the gap between antibody-based detection and discovery-based mass spectrometry.

In this book, a panel of high-profile authors present state-of-the-art targeted proteomics applications in biomedical areas, a field that is advancing rapidly on many fronts. The first part of the book provides examples of the application of targeted proteomics approaches in biomarker discovery and pathogenesis. The second part is focused on different tools which currently are useful in targeted proteomics, such as bioinformatics approaches. The last part is dedicated to biobanks, which are a critical aspect in biomedical research because the quality of the data from targeted proteomics relies entirely on the quality of the samples used to generate the data.

Manuel Fuentes and Joshua LaBaer

Serum Proteomics for Studying Disease Pathogenesis and Identification of Disease Biomarkers

1

Parvez Syed, Sandipan Ray, Kishore Gollapalli and Sanjeeva Srivastava

Abstract

It is important to understand the disease pathogenesis and host immune response to make a correct therapeutic decision. Most of the diseases manifest at the protein level and hence analysing changes at protein level gives an insight into the pathophysiological condition of the patients. However, identifying such subtle differences in protein expression is not an easy task. Over the decades, various proteomic techniques have been developed to perform serum proteome analysis. In this chapter we discuss the application of gel-, mass spectrometry- and array-based methods for the serum proteome analysis and identification of protein biomarkers. Advances in proteomic technologies have enhanced the diagnostic potential of existing routine clinical approaches and help in assessing the risk factors. We also discuss the challenges associated with serum proteome profiling using these proteomic approaches.

Introduction

Genomic studies provide a wealth of information; however, they do not give us a detailed understanding of the biomolecules, such as proteins, which govern the cellular processes (Ali-Khan *et al.*, 2003). Often it has been observed that abundance of DNA or RNA is not a precise indicator of protein abundance. Proteomics aims to fill the gap between the genomic information and functional proteins (Tan *et al.*, 2012; Vogel *et al.*, 2010). Identification and quantification of such proteins help in understanding the molecular dysfunctions associated with the progression of diseases (Anderson, 2010). For the identification of such biomarkers, proteomics along with genomics and bioinformatics can play a crucial role (He and Chiu, 2003). Proteomics studies can be performed on various body fluids such as blood (serum or plasma), cerebrospinal fluid, urine or saliva. Proteomics is relatively new field of science; however, the concept of identification of biomarkers in biological fluids is very old. For example, Bence Jones protein was identified as cancer marker in urine in 1847. Since then many techniques, such as electrophoresis, chromatography, immunoassays and serological assays, have been developed for qualitative and quantitative analysis of hundreds of proteins in biological fluids (Hortin *et al.*, 2006).

The proteins present in the body fluids represent the physiological and pathophysiological condition of a disease and can be very useful in diagnosing the disease in the early stages so that a proper therapeutic decision can be made (Schiffer *et al.*, 2006). Although obtaining saliva is the easiest method and non-invasive, the amounts of informative analytes are lower than in other body fluids (Zhang *et al.*, 2013). Because of the reach of the circulatory system into various tissues and organs, blood is very rich in information about pathophysiological condition. The need to find protein markers to enable diagnosis of diseases at the early stages have made serum/plasma the most studied body fluid using proteomics (Issaq *et al.*, 2007). In this chapter we discuss various proteomic approaches for identifying changes in the protein expression levels in serum and the challenges associated with serum proteomic analysis.

Challenges of serum proteome analysis

The comprehensive proteomic analyses of different biological fluids, particularly serum and plasma, have attracted considerable interest for the identification of diagnostic and prognostic protein biomarkers (Ray *et al.*, 2011). Serum/plasma is considered an attractive sample for clinical research, since it is easily accessible and contains different types of proteins released by various diseased tissues and provides a comprehensive representation of the pathophysiological condition of a patient (Anderson and Anderson, 2002). Serum proteome responds

Box 1.1 Challenges of serum proteomic analysis

- Pre-analytical variations are introduced during sample collection, handling and storage process. Serum/plasma collection devices, separation process and storage temperature are crucial factors which considerably affect the sensitivity, selectivity and reproducibility of the down-stream proteomics analysis.
- Extremely broad dynamic range of protein concentrations in serum/plasma. Serum/plasma proteome is very complex due to the presence of huge diversity of proteins in a very dynamic concentration (10^{10}–10^{12}), which is beyond the range of majority of the conventional gel/MS-based proteomic techniques.
- Masking effect due to the presence of high-abundance proteins. Eventually, 22 most abundant proteins represent about 99% of the total protein mass in plasma, where only albumin covers approximately 50%. Consequently, it is very challenging to identify the low-abundance potential markers, unless the high-abundance proteins are removed or extensive pre-fractionation approaches are applied.
- High levels of salts and other interfering compounds. The presence of excessive salts and other contaminants in blood often badly affects the downstream experimental procedure and prevents accurate separation and identification of target proteins.
- Fragile nature of proteins. Often the marker proteins are very unstable and degraded during the analysis process.
- Presence of various isoforms of single protein due to post-translational modifications (PTMs). Biological activities and molecular functions of majority of the eukaryotic proteins are controlled by PTMs. Efficient and sensitive analysis of PTMs is quite challenging.
- Huge variation in the proteome with time within a same individual. The proteome of an organism is highly dynamic and altered with time so analysis of temporal samples and designing of longitudinal studies are very important.
- Differences among the individuals from the same or different populations. Due to the existence of variations among individuals depending on sex, age, genetic factors, dietary considerations, environmental factors and drug treatment, much bigger clinical cohorts are crucial for the establishment of 'gold-standard' biomarkers.

rapidly to external stimuli and disease, and the correlation between serum protein levels and disease progression is of particular interest from clinical and prognostic perspectives. Over the last decade diagnostic applications of serum/plasma proteomics have been growing steadily and established as a promising approach for the identification of surrogate proteins markers and studying disease pathogenesis and host immune responses in cancer and other fatal human diseases (Anderson, 2010; Ray *et al.*, 2011).

Biomarkers are indicator biomolecules that assist in early diagnosis, discriminate between different diseases, and are valuable for monitoring progression/severity of disease. Despite different advancements, even now there are several biological and technological limitations for the current proteomics technologies, which are frequently applied for discovery of disease related marker proteins in serum/plasma samples (Issaq *et al.*, 2007). Such limitations are described in detail in Box 1.1. In brief, the pre-analytical variations introduced during sample collection, handling and storage process, are also challenging for screening true biomarkers (Rai *et al.*, 2005). In addition, because of the very wide dynamic range of protein concentrations and presence of high-abundance proteins, low-abundance marker proteins are masked. Other factors which hinder the direct application of proteomic technologies in clinics are the high content of salts along with other interfering components in most of the biological specimens.

Information obtained from the proteome level analysis of blood and other types of bio-fluids are very promising for translational research and can contribute significantly in unravelling various unexplored mysteries of human diseases and enlighten the way of proteomics in different applications of clinical research; including identification of new drug and vaccine targets, establishment of diagnostic and prognostic biomarkers (Omenn, 2004). Nevertheless, different technological limitations associated with the existing proteomics approaches are holding back the bedside implications of proteomics approaches. Imminent future of this highly promising research field is dependent on the successful resolution of the existing limitations. There is an urgent need for collaborative initiatives at a global level to prepare standard protocols for clinical proteomics research to bring uniformity in sample collection, storage, processing and data analysis procedures, to avoid pre- and post-analytical variations.

Proteomic techniques in serum profiling

Gel-based techniques

Sodium dodecyl sulfate polyacrylamide gel electrophoresis

Sodium dodecyl sulfate polyacrylamide gel electrophoresis or SDS-PAGE is one of the most commonly used single-dimension protein separation methods. The denatured proteins are separated in polyacrylamide gel electrophoresis (PAGE) on the basis of their molecular weight. However, this method is inefficient in separating complex protein extracts (Rath *et al.*, 2009; Rabilloud *et al.*, 2010). Proteins of the same molecular mass cannot be separated using this technique. To overcome this limitation, two-dimensional gel electrophoresis was developed, which can separate proteins with same masses but different isoelectric points (pI).

Two-dimensional electrophoresis

To separate complex protein extracts, a strategy based on molecular weight has to be complemented with another separation technique such as isoelectric focusing (IEF) (Rabilloud *et al.*, 2010). The concept of two-dimensional electrophoresis (2-DE) was independently developed by Klose (1975) and O'Farrell (1975). In 2-DE, the IEF is performed using immobilized pH gradient (IPG) strips (Bjellqvist *et al.*, 1982) (Fig. 1.1A).

Two-dimensional electrophoresis, despite being one of the most widely used methods in the field of proteomics, suffers from a few limitations. The amount of proteins that can be loaded onto 2-DE gels is insufficient to allow proper characterization of the proteins, and only soluble proteins can be studied using 2-DE. However, most of the regulatory proteins are low abundance proteins and such low abundance proteins are overwhelmed by the presence of higher amount of abundant proteins. To address these issues, sample pre-fractionation is done followed by separation on narrow range IPG strips. Despite these measures, low-abundance proteins, hydrophobic or proteins with pI > 10 and proteins with extreme sizes cannot be studied properly using 2-DE (Harry *et al.*, 2000; Lescuyer *et al.*, 2004). The biggest drawback of 2-DE is the reproducibility and gel-to-gel variations since the different sample types are run on different gels (Becnel and McKenna, 2012). The sensitivity of this method depends on the agent used to stain the protein spots. Detection of very low-abundance proteins requires very sensitive fluorescent dyes, which is not possible using classical 2-DE method.

Two-dimensional difference in gel electrophoresis

Two-dimensional difference in gel electrophoresis (2D-DIGE), a modified and advanced version of 2-DE, was developed by Unlü *et al.* (1997). They used two different fluorescent dyes (Cy3 and Cy5) for two different samples and ran them on the same gel. The gel images were obtained by exciting the dyes at their corresponding wavelengths and then these two images were superimposed to observe the differentially expressed proteins. In 2D-DIGE, a complex mixture of proteins from different sample types is separated on the same gel based on the molecular mass and charge. Running different sample types on the same gel eliminates the gel-to-gel variation and thus avoids the artefacts owing to the sensitive nature of the Cy-dyes used in 2D-DIGE, the visualization of low-abundance proteins is possible (Becnel and McKenna, 2012; Kosako *et al.*, 2009; Dubrovska and Souchelnytskyi, 2005). Using

Figure 1.1 Schematics of different proteomic techniques used in serum proteome profiling for biomarker discovery. (A) Gel-based proteomics. Gel-based protein profiling techniques like 2-DE involve multiple serum processing steps including depletion of high-abundance proteins and removal of impurities. In 2-DE separation of proteins are based on two parameters: isoelectric points (first dimension) and relative molecular mass (second dimension). (B) MS-based proteomics using MALDI-TOF-TOF: MALDI is an efficient process for generating gas-phase ions of peptides and proteins for mass spectrometric detection. Accurate determination of the mass of the protein species present in the sample is achieved by the use of a time-of-flight (TOF) mass analyser, which resolves ions based on *m/z* ratios. The first tube separates the ions generated by laser beam and the second one resolves the fragmented species generated by a collision chamber, which is present between the two TOF tubes. (C) Array-based proteomics; protein microarrays, where proteins of interest are printed on the solid support and the biological sample under study is dispensed on to the array surface, which leads to selective binding of target proteins. Detection of the target proteins is performed using labelled secondary molecules. ▷

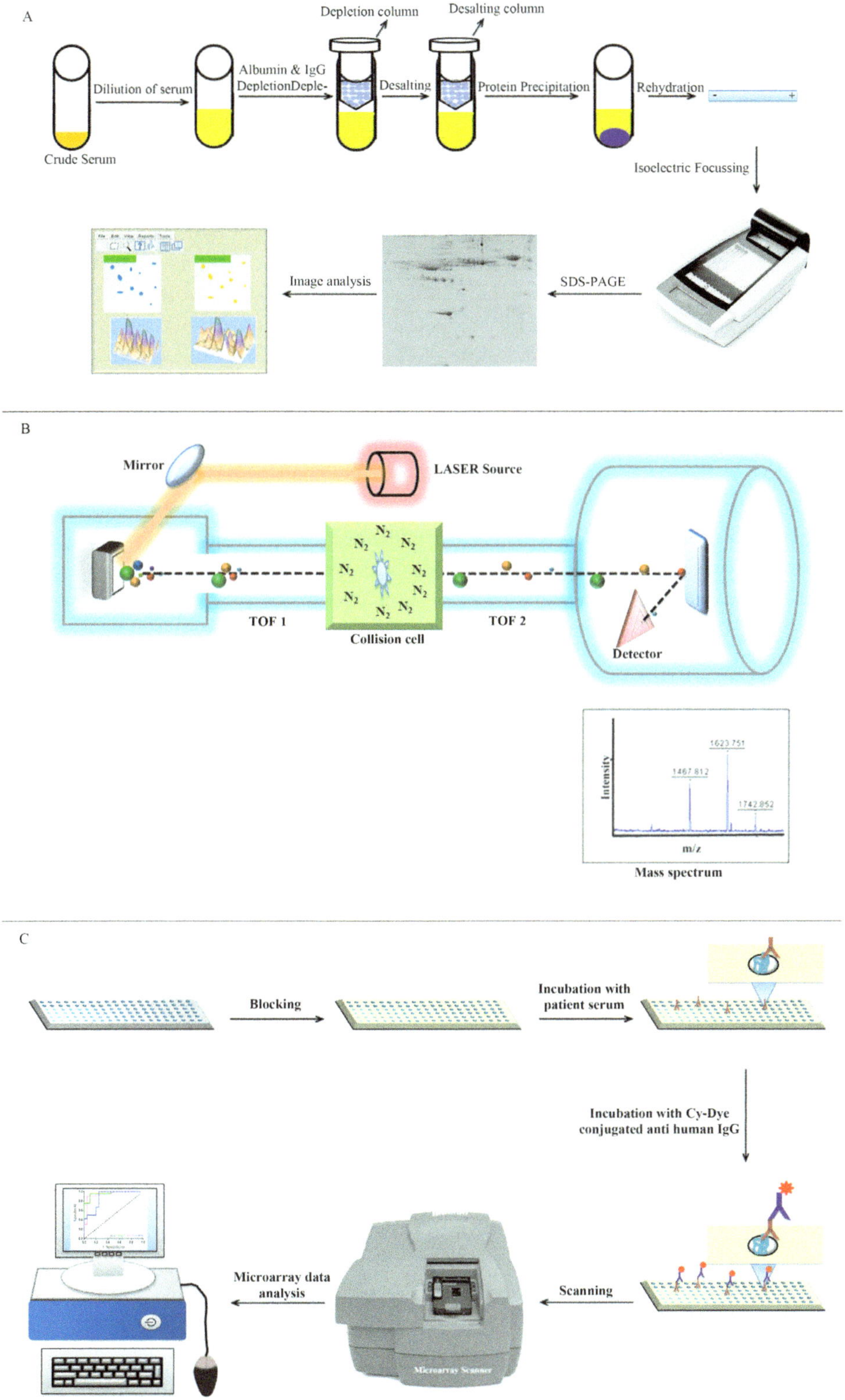
A
Depletion column
Desalting column
Diliution of serum
Albumin & IgG
DepletionDeple-
Desalting
Protein Precipitation
Rehydration
Crude Serum
Isoelectric Focussing
Image analysis
SDS-PAGE
B
Mirror
LASER Source
TOF 1
Collision cell
TOF 2
Detector
Intensity
1623.751
1467.812
1742.852
m/z
Mass spectrum
C
Blocking
Incubation with patient serum
Incubation with Cy-Dye conjugated anti human IgG
Microarray data analysis
Scanning
Microarray Scanner

pre-electrophoresis Cy-dye labelling or employing sensitive post-electrophoresis staining agents, the low-abundance proteins can be visualized; however, identification of those low-abundance proteins remains difficult owing to the unavailability of sufficient peptides during the in-gel digestion process, which is essential for subsequent mass-spectrometry based protein identification.

Mass spectrometry-based techniques

In the early days of mass spectrometry-based (MS-based) techniques, the available DNA sequence information was used along with the mass spectrometric molecular weight information for the identification of proteins and peptides using fast atom bombardment and plasma desorption mass spectrometry (Roepstorff, 2012). The advent of techniques such as matrix-assisted laser desorption/ionization (MALDI) (Karas and Hillenkamp, 1988) and electrospray ionization (ESI) (Fenn *et al.*, 1989) made the identification of tryptic digested proteins based on the molecular mass of the peptides possible (Mann *et al.*, 1993; James *et al.*, 1993; Pappin *et al.*, 1993). However, identification of tryptic peptides was not sufficient enough for the confident identification of proteins. To address this issue, in 1994, Mann and Wilm came up with the concept of sequence tags. In this method, they combined the molecular mass information with the partial sequence information obtained from MS/MS for more confident protein identification (Mann and Wilm, 1994). For the ionization of the protein samples, there is another method known as surface-enhanced laser desorption/ionization (SELDI), which is a variation of laser desorption/ionization (LDI) developed by Hutchens and Yip in 1993 (Hutchens and Yip, 1993). SELDI in combination with mass spectrometer like time-of-flight (TOF) has been extensively used for the identification of serum biomarkers in cancers (Laronga *et al.*, 2003; Garrisi *et al.*, 2008). Using SELDI-TOF MS for profiling sera obtained from breast cancer patients, Opstal-van Winden *et al.* (2011) showed that altered protein profiles in the serum can be detected up to 3 years prior to the onset of the disease. Similarly, Cheng *et al.* (2008) also used SELDI-TOF MS for serum profiling of laryngeal squamous cell carcinoma and found a panel of biomarkers which can differentiate various grades of the disease from one another with high sensitivity and specificity.

Unlike gel-based methods, LC-MS based strategy facilitates the identification of proteins without having the need to separate the proteins. Typically, the proteins are tryptic digested and subjected to separation using liquid chromatography. In order to analyse complex mixture of thousands of proteins, one needs to perform more than one chromatography such as ion exchange chromatography (e.g. strong cation exchange (SCX) followed by reverse phase-high-performance liquid chromatography (RP-HPLC)). Once the tryptic digested peptides are separated using LC, ionization of the peptides is done using ESI followed by analysis using mass analysers (Fig. 1.1B). There are two types of mass analysers such as scanning and ion-beam mass spectrometers (e.g. TOF and Q) and trapping mass spectrometers [e.g. IT, Orbitrap and Fourier transform ion cyclotron resonance (FT-ICR)] (Roepstorff, 2012; Yates *et al.*, 2009). Using a highly sensitive mass spectrometer interfaced with a nanoLC could be useful for detection of very low-abundance proteins. Recently, Robinson's group from the University of Pittsburgh, USA, identified 689 proteins from the human plasma samples using LTQ-Orbitrap Velos mass spectrometer (Thermo Fisher Scientific, Waltham, MA, USA) (Cao *et al.*, 2013).

Protein microarrays

For studying interactions between proteins and peptides in a high-throughput manner, protein microarrays serve as an excellent platform and can be compared to immunoassays such as radioimmunoassay (RIA) and ELISA. Prior to the advent of protein microarrays, macroarrays spotted with cDNA expression clones were used for such studies. The recombinant proteins expressed from these expression clones are directly immobilized on the array surface and these recombinant proteins are used for serum profiling (Fig. 1.1C) (Luna Coronell *et al.*, 2012; Stempfer *et al.*, 2010). As early as 1990, Ekins *et al.* (1990) suggested that a solid surface of few microns can yield results sensitive enough to be compared to that of macroarrays. However, the dynamic range of the signal intensities derived from macroarrays are inferior to the 16 bit ($0–2^{16}$) dynamic range of protein microarray (Stempfer *et al.*, 2010).

The advancements in the field of DNA microarrays have paved the way for the development of protein microarrays. Since then protein microarrays have been extensively used for cancer biomarker discovery (Luna Coronell *et al.*, 2012). Prior knowledge of the identity of the identified protein gives protein microarray technology an upper-hand over the other proteomic techniques. Such prior knowledge would immensely help in efficient validation of the potential biomarkers using classical techniques like Western and dot blots, ELISA and immunohistochemistry (IHC) (Kalinina *et al.*, 2011). The protein microarrays can be categorized into different types, like analytical, reverse phase and functional protein microarrays, based on their purpose and the molecules immobilized on the surface of the protein microarrays. Analytical protein microarrays apart from enabling profiling of complex protein mixtures can also be used for the measurement of binding affinities, specificities and protein expression levels. The analytical protein microarrays are arrayed with antibodies as capturing molecules. Reverse-phase protein microarrays are spotted with lysate protein extracts from various tissues and probed with labelled antibodies. Similarly, the microarrays where the immobilized capturing molecules on a microarray surface are antibodies are known as forward phase protein microarrays. The detection of the interaction is done using fluorescently tagged proteins. This approach could be problematic, as there is a possibility of having those tags within the epitope regions and this might lead to the steric interference of antibody binding. Yet another type of protein microarrays is the functional protein microarrays and these microarrays are used to study the biochemical aspect of an entire proteome in a single experiment (Hall *et al.*, 2007; Phizicky *et al.*, 2003).

Based on the surface the capture molecules are immobilized, there are two kinds of protein microarrays, planar and bead-based protein microarrays. Planar protein microarray is solid-phase assay system which contains thousands of capture molecules immobilized on a coated planar surface. The typical diameter of a microspot on a microarray is $<250\,\mu m$. The immobilization of the capturing molecules is made possible by applying various surface chemistry strategies. Making the surface suitable for capture molecule immobilization leads to the problems associated with non-specific binding during the assay, especially because serum contains a lot of proteins. It is extremely important, prior to the assay, to block the surface of the microarrays and by doing so one can achieve high signal-to-noise ratio (Hartmann *et al.*, 2009; Zhu and Snyder, 2003; Angenendt *et al.*, 2003; Seurynck-Servoss *et al.*, 2007). There are two types of assays one can perform using planar protein microarrays, one is label-based assay and another is sandwich assay. Label-based assays are also known as direct labelling assays, where fluorophores such as Cy3 and Cy5 are tagged to an antibody or antigen and this labelled protein is applied to a microarray containing antibody or antigen on its surface. However, direct-labelling method has limited sensitivity and specificity. In sandwich assay, otherwise known as indirect-labelling assay, involves capturing of unlabelled proteins by the immobilized capturing proteins on the microarray surface. A detection antibody labelled with a fluorescent dye is used for the detection of unlabelled protein. To arrive at a meaningful conclusion, the fluorescent signal intensities are measured using scanners and statistical analysis is performed (Sanchez-Carbayo, 2006; Hall *et al.*, 2007).

On the other hand, in bead arrays, each fluorescently labelled bead is coated with a different capture antibody. The suspension of the sample is added to the antibody coated fluorescent beads and this allows the antibody–antigen interaction. The suspension along with the beads is mixed with a cocktail of fluorescently tagged detection antibodies. This cocktail of detection antibodies contains antibodies specific to each one of the capture antibodies. Using a flow cytometer system, the amounts of the detection antibodies on the beads are determined (Sanchez-Carbayo, 2010; Lash *et al.*, 2006; de Jager and Rijkers, 2006; Waterboer *et al.*, 2006).

Nanoproteomics

While proteomics has emerged in different related disciplines of biological sciences, other technologies are also incorporated in proteomics research to advance the capabilities of existing proteomics techniques and overcome their shortcomings. The amalgamation of proteomics with nanotechnology has led to the innovation of a new interdisciplinary platform called 'nanoproteomics'. This interdisciplinary endeavour places a strong emphasis on improving the sensitivity and multiplexing capability of routinely used proteomic techniques (Ray *et al.*, 2010, 2011). During the last decade, several nanostructured materials, such as quantum particles, carbon nanotubes, silicon nanowires and gold nanoparticles, have been incorporated in proteomics research. These components have become popular among the research communities because of their high compatibility with MS-based or array-based proteomics. Nanoproteomics offers multiple advantages, which include ultra-high sensitivity, fast detection and suitability for multiplexing.

SEREX

Aberrant or mutated protein expression evokes human immune response and as a result these proteins are recognized as self-antigens or autoantigens. Such autoantigens have been identified in many autoimmune diseases and cancers using serological identification of antigens by recombinant expression cloning (SEREX) developed by Pfreundschuh and colleagues (Sahin *et al.*, 1995, 1997). SEREX involves construction of cDNA expression libraries from patients' tissue. These libraries are cloned into *E. coli* and the recombinant proteins are expressed by applying a nitrocellulose membrane soaked in isopropyl-β-D-1-thiogalactopyranoside (IPTG) and the expressed proteins are tethered to the nitrocellulose membranes. The serum pools from the patients are applied onto these membranes and the detection is performed using horseradish peroxidase (HRP)-conjugated antibody. The corresponding colonies are picked from the culture plates and sequenced for the identification of the expressed proteins and such proteins can be used further for validation using microarrays (Syed, 2012). By using the proteins from the SEREX clones derived from brain and cancer tissues on protein microarrays, Stempfer *et al.* (2010) could differentiate the serum samples from each other.

Applications

Based on the applications, proteomics can be broadly classified into three categories, protein expression proteomics, structural proteomics and functional proteomics (He and Chiu, 2003). The advances made in the last decades in the field of proteomics enabled the high-throughput identification of proteins. These approaches, compared to the traditional single-protein approach, help in better understanding biological mechanisms (Abbott, 1999). The protein expression proteomics deals with studying different proteins expressed in a given biospecimen, such as tissue, biological fluid or cell lysates, with the aim of identifying biomarkers or drug targets. In functional proteomics, PTMs, protein–DNA/RNA or protein–protein interactions of a group of proteins falling under a specific biological function are studied. The third category of proteomic studies, structural proteomics, deals with structural aspects of the proteins in a specific cell organelle (He and Chiu, 2003). In this section, we discuss in detail how the protein expression based proteomics can be applied for the identification of biomarkers in cancers, infectious and autoimmune diseases.

Cancers

Uncontrolled proliferation of cells results in the cancer development. According to the International Agency for Research on Cancer (IARC) statistics for 2008, a total of 7.6 million people were died from cancers and 12.7 million cancer cases were registered all around the world. Most cancer deaths occur because the disease is only detected at a very advanced stage. The sophisticated tools/diagnostic methods for the early detection of cancers are lacking, especially in developing countries (Jemal *et al.*, 2011). Serum is one of the promising biological fluids for the discovery of cancer biomarkers. In most cancers, cancer tissues release proteins into the bloodstream either by secretion or on cell death (Zhang *et al.*, 2007). Identification of these proteomic alterations could have diagnostic/prognostic/therapeutic value. Because of its ease of collection, serum has become one of the most attractive sources for the biomarker discovery (Ray *et al.*, 2011). Proteomics is being used in the discovery of serum biomarkers for various cancers such as prostate cancer, ovarian cancer, brain cancer, breast cancer and pancreatic cancer, etc. (Soydinc *et al.*, 2012; Khan *et al.*, 2012; Bast, Jr. *et al.*, 1983; Kumar *et al.*, 2010; Gollapalli *et al.*, 2012; Poruk *et al.*, 2013). Significant work is going on all around the world for the discovery of early diagnostic markers for cancers. In most of the other diseases, including cancers, we can see the increased serum levels of acute phase proteins in response to inflammatory reactions (Gruys *et al.*, 2005). Changes in the serum/plasma levels of various proteins like survivin in the prostate cancer (Khan *et al.*, 2012), cancer antigen 125 (CA-125) levels in the ovarian cancer (Bast Jr. *et al.*, 1983), haptoglobin α2, retinol-binding protein in glioblastoma multiforme (Kumar *et al.*, 2010; Gollapalli *et al.*, 2012), urokinase-type plaminogen activator in the breast cancer (Soydinc *et al.*, 2012), serum osteopontin and tissue inhibitor of metalloproteinase 1 in pancreatic cancer (Poruk *et al.*, 2013) were already reported in the literature.

Mutations are common in most of the cancers. These mutations or abnormal processing of proteins may change the native structure of the proteins resulting in the generation of auto antibodies against these proteins (Wandall *et al.*, 2010). In many cancers autoantibodies can be detected even before the on-set of the disease. So autoantibodies might be good diagnostic markers for cancers. Autoantibodies for the extracellular protein kinase-A (ECPKA) are found in many cancers (Nesterova *et al.*, 2006). Autoantibodies generated against O-glycoforms of extracellular mucin also found in breast, ovarian and prostate cancer (Wandall *et al.*, 2010). Further, the combination of altered serum protein markers along with the autoantibodies increases the specificity and sensitivity for the disease diagnosis (Dong *et al.*, 2013).

Glioblastoma multiforme (GBM) is a type of brain cancer which originates from the glial cells. Average survival period of these patients is 12–14 months after diagnosis. Diagnosis of this cancer, before on-set of this disease might be useful in the proper treatment of these patients. Significant work has been done in the field of serum proteomics of glioblastomas. Various research groups from India are working on this cancer. Our study on the GBM patients' serum samples using gel based proteomic approach showed the differential expression of 28 proteins in the GBM patients with respect to the healthy subjects. Serum proteins like ceruloplasmin, vitamin D-binding protein, serum amyloid P showed good correlation with the cancer growth and development. Various physiological pathways were found to be altered in the GBM patients like plasminogen activating cascade, vitamin D metabolism, phosphatidylinositol 3-kinase signalling pathway, etc. A subset of the differentially expressed protein data were used for the development of class prediction models, which were able to discriminate the GBM population from the healthy subjects. The significantly altered serum proteins in our study could be potential candidate markers for the disease diagnosis and understanding the disease pathobiology (Gollapalli *et al.*, 2012).

Serum/plasma proteomics of infectious diseases

Existing and re-emerging infectious diseases are regarded as important global public health problems and major causes of disease-related mortalities, especially in developing countries. Although direct detection of infecting organisms or pathogen markers is regularly used for diagnosis of infections in clinics, surrogate host markers can act as an effective alternative for sensitive and early diagnosis of pathogenic infections (TDR Diagnostics Evaluation Expert Panel *et al.*, 2010). Blood biomarkers for pathogenic infections can be used for early diagnosis, discrimination between closely related infections with similar clinical manifestations as well as aid in monitoring disease progression, response to therapy, and predicting outcomes. In addition, investigation of the pathogen induced alterations in host proteome under diseased conditions can provide valuable information regarding disease pathogenesis and host immune responses (Ray *et al.*, 2011). Consequently, over the last decade huge numbers of proteomics studies have been conducted to identify differentially expressed serum/plasma proteins in various types of infectious human diseases including viral infections such as severe acute respiratory syndrome (Chen *et al.*, 2004), and swine fever virus infection (Sun *et al.*, 2011), parasitic infections like malaria (Kassa *et al.*, 2011), bacterial infections such as tuberculosis (Tanaka *et al.*, 2011) with an aspiration of discovery of novel potential diagnostic/prognostic biomarkers and/or deciphering mysteries of disease pathobiology. Nevertheless, critical review of literature indicates the expression levels of many of the reported potential biomarkers are commonly altered in multiple infectious diseases regardless of the type of infection, which might be a consequence of inflammation mediated acute phase response signalling, while the uniquely modulated candidates in each pathogenic infection could be the indicators of some inimitable responses. Our research group is also investigating the alterations in human serum proteome in different tropical infectious diseases including *falciparum* and *vivax* malaria (Ray *et al.*, 2012a,b), dengue fever (Ray *et al.*, 2012c) and leptospirosis (Srivastava *et al.*, 2012) to categorize the common and distinctive signatures.

As mentioned previously, there are several published articles demonstrating the application of serum proteomics analysis to the study of different infectious diseases. Here, we have discussed the implication of gel and MS-based proteomics for identification of surrogate markers for malaria. Malaria is the most prevalent tropical parasitic disease with a worldwide incidence of 300–500 million clinical cases per year and is the cause of at least one million deaths yearly. We have analysed alterations in the human serum proteome as a result of infection by malaria parasites *Plasmodium falciparum* and *P. vivax*. These analyses have given us the mechanistic insights about disease pathogenesis, host immune response and has led to the identification of potential protein markers in Indian populations using two complementary quantitative proteomic approaches; difference in-gel electrophoresis (2D-DIGE) and isobaric tag for relative and absolute quantitation (iTRAQ). By using comparative serum proteomic analysis approaches we have identified several differentially expressed proteins and modulation of associated vital physiological pathways in *P. falciparum* and *P. vivax* infected patients compared to the healthy subjects (Ray *et al.*, 2012a,b). Although quite a few serum proteins were found to be commonly altered in other different infectious diseases such as dengue fever and leptospirosis and might be a consequence of inflammation mediated acute phase response signalling, uniquely modulated candidates were also identified in each pathogenic infection indicating some unique responses. Interestingly, a panel of our identified serum proteins (serum amyloid A, haemopexin, apolipoprotein E, haptoglobin, retinol-binding protein and apolipoprotein A-I) has shown the ability to discriminate *vivax* malaria from healthy subjects and patients suffering from other non-malarial infectious diseases. The accurate discrimination among the different infectious disease groups on the basis of differentially expressed proteins demonstrates the potential diagnostic implications of such proteomics approaches.

As is frequently experienced with several proteomics biomarker discovery studies, extremely exciting and promising findings identified in smaller populations did not live up to the expectations at the validation stages and subsequent larger clinical trials. Therefore, designing of appropriate proteomics experimental set-up involving different types of febrile disease controls and validation of the initial findings in bigger clinical cohorts using alternative technological approaches is essential for obtaining any conclusive protein markers for any specific infection, which can be translated further in actual clinical applications.

Autoimmune diseases

Autoimmune diseases are the result of host immune system targeting self proteins or autoantigens and serum proteome analysis can be performed to identify the autoantibodies involved (Hueber and Robinson, 2006). The role of the autoantigens in autoimmune diseases have been observed in patients suffering with variety of autoimmune diseases like Sjögren's syndrome, systemic lupus erythematosus, rheumatoid arthritis (RA) and multiple sclerosis (Robinson *et al.*, 2002; Schett and Gravallese, 2012; Mayer *et al.*, 2012). For identifying aberrantly expressed proteins in the body fluids from the patients suffering with autoimmune diseases, gel-based, MS-based methods and protein microarrays have been employed by different research groups (Fathman *et al.*, 2005). In this section we will briefly discuss autoantibody studies in multiple sclerosis and RA.

Multiple sclerosis is a neurodegenerative disorder whose pathogenesis and aetiology is poorly understood. In the early stages of the disease, the autoreactive lymphocytes migrate through the blood–brain barrier into the central nervous system and as a result the myelin sheath wrapping the axons becomes damaged (Fathman *et al.*, 2005; Compston and Coles, 2002). Although, there are many potential biomarkers from various serum profiling studies, very few are clinically validated. So far, the search for multiple sclerosis-specific serum biomarkers seems illusive and this can be attributed to the heterogeneous nature of the disease. Additionally, the autoantibodies identified to date vary in the roles they play in the disease progression (Graber and Dhib-Jalbut, 2011; Meinl *et al.*, 2006).

Rheumatoid arthritis is another autoimmune disease whose aetiology is affected by genetic and environmental factors. Patients with RA suffer with inflammation of multiple joints and bone erosion (Scott *et al.*, 2010; Ortea *et al.*, 2012). An early autoimmune response observed in RA triggers the production of autoantibodies, and the presence of these autoantibodies in the serum can be used to develop diagnostic tools for detecting RA in its early stages (Nell *et al.*, 2005). The most common prognostic markers observed in the sera of RA patients are IgM and IgA rheumatoid factors against the Fc region of IgG. Yet other important RA markers, observed in most of the RA patients, are the antibodies against citrullinated peptides (ACPA) (van der Linden *et al.*, 2009). Using gel-based and MS-based proteomics, Goëb *et al.* (2009) identified two classes of antigens, citrullinated glycolytic enzymes and molecular chaperones, which evoke autoantibody response in RA patients. Protein microarrays have also been instrumental in identifying autoantibodies in the serum of RA patients (Szodoray and Alex, 2011).

Conclusion

Minimal invasiveness, easy collection methods and close correlations with diseases states make serum an attractive biospecimen for the identification of disease biomarkers. By using the serum proteome approach one can avoid unnecessary biopsies. The serum acts as a reservoir of potential biomarkers because of the spread of the circulatory system in all the tissues and the organs (Tan *et al.*, 2012). In cancers, tumour autoantibodies in the patient sera can be identified prior to the symptoms (Tan *et al.*, 2009). These properties make serum an important specimen to identify potential biomarkers and help us to get an insight into the host's response to various diseases. Serum samples can be used not only for diagnostic purposes but also for risk prediction, recurrence of disease, response to drugs, etc. To perform such studies, one needs to ensure proper collection methodology and storage facility. Proper records of the blood samples need to be maintained by the biorepositories, which would enable a proper follow-up study and evaluating the risk.

Once the serum samples are obtained, one can use gel-based, MS-based or array-based proteomics for the identification of disease biomarkers. Each technique has its own sets of advantages over another. MS-based techniques give a comprehensive coverage of the proteins expressed in the patients and helps in identification of the differentially expressed proteins. However, for analysis of complex samples like serum, one needs to perform pre-fractionations to reduce the complexity of the sample. Protein microarrays serve as an excellent alternative, since the serum which needs to be analysed need not to be depleted.

Often the differentially expressed proteins identified by various proteomic techniques might not be the result of the disease manifestation and just mere bystanders. For the identification of biomarkers that can be clinically useful, there is a need for analysing larger cohorts. Once a robust panel of biomarkers are derived from these proteomic techniques, the validation should be performed using rather classical techniques, like ELISA, RIA, etc., which can be performed easily at any clinical settings without requiring highly skilled personnel or sophisticated instrumentations. Proteomic techniques have been evolving and will continue to evolve and find innovative ways of discovering very specific biomarkers enabling better diagnosis, prognosis and risk assessment. Due to various technical advancements the possibility of identifying and validating the serum/plasma biomarkers for different human diseases is now higher than ever before. Human Plasma Proteome Project (HPPP) is one of the Human Proteome Organization's (HUPO) initiatives, which aim to study and analyse human plasma/serum specimens. The eventual ambition of clinical proteomics is to translate the finding of basic research into direct practical clinical applications through expansion of processes/products, which can effectively advance diagnostics and therapeutics.

References

Abbott, A. (1999). A post-genomic challenge: learning to read patterns of protein synthesis. Nature *402*, 715–720.

Ali-Khan, N., Zuo, X., and Speicher, D.W. (2003). Overview of proteome analysis. Curr. Protoc. Protein Sci. Chapter 22, Unit 22.1.

Anderson, N.L. (2010). The clinical plasma proteome: a survey of clinical assays for proteins in plasma and serum. Clin. Chem. *56*, 177–185.

Anderson, N.L., and Anderson, N.G. (2002). The human plasma proteome: history, character, and diagnostic prospects. Mol. Cell Proteomics. *1*, 845–867.

Angenendt, P., Glokler, J., Sobek, J., Lehrach, H., and Cahill, D.J. (2003). Next generation of protein microarray support materials: evaluation for protein and antibody microarray applications. J. Chromatogr. A *1009*, 97–104.

Bast, R.C., Jr., Klug, T.L., St John, E., Jenison, E., Niloff, J.M., Lazarus, H., Berkowitz, R.S., Leavitt, T., Griffiths, C.T., Parker,L., *et al.* (1983). A radioimmunoassay using a monoclonal antibody to monitor the course of epithelial ovarian cancer. N. Engl. J. Med. *309*, 883–887.

Becnel, L.B., and McKenna, N.J. (2012). Minireview: progress and challenges in proteomics data management, sharing, and integration. Mol. Endocrinol. *26*, 1660–1674.

Bjellqvist, B., Ek, K., Righetti, P.G., Gianazza, E., Gorg, A., Westermeier, R., and Postel,W. (1982). Isoelectric focusing in immobilized pH gradients: principle, methodology and some applications. J. Biochem. Biophys. Methods *6*, 317–339.

Cao, Z., Yende, S., Kellum, J.A., and Robinson,R.A. (2013). Additions to the Human Plasma Proteome via a Tandem MARS Depletion iTRAQ-Based Workflow. Int. J. Proteomics. 2013, 654356.

Chen, J.H., Chang, Y.W., Yao, C.W., Chiueh, T.S., Huang, S.C., Chien, K.Y., Chen, A., Chang, F.Y., Wong, C.H., and Chen, Y.J. (2004). Plasma proteome of severe acute respiratory syndrome analyzed by two-dimensional gel electrophoresis and mass spectrometry. Proc. Natl. Acad. Sci. U.S.A. *101*, 17039–17044.

Cheng, L., Zhou, L., Tao, L., Zhang, M., Cui, J., and Li, Y. (2008). SELDI-TOF MS profiling of serum for detection of laryngeal squamous cell carcinoma and the progression to lymph node metastasis. J. Cancer Res. Clin. Oncol. *134*, 769–776.

Compston, A., and Coles, A. (2002). Multiple sclerosis. Lancet *359*, 1221–1231.

Dong, X., Yang, M., Sun, H., Lu, J., Zheng, Z., Li, Z., and Zhong, L. (2013). Combined measurement of CA 15–3 with novel autoantibodies improves diagnostic accuracy for breast cancer. Onco. Targets. Ther. *6*, 273–279.

Dubrovska, A., and Souchelnytskyi, S. (2005). Efficient enrichment of intact phosphorylated proteins by modified immobilized metal-affinity chromatography. Proteomics. *5*, 4678–4683.

Ekins, R., Chu, F., and Biggart, E. (1990). Multispot, multianalyte, immunoassay. Ann. Biol. Clin. (Paris) *48*, 655–666.

Fathman, C.G., Soares, L., Chan, S.M., and Utz, P.J. (2005). An array of possibilities for the study of autoimmunity. Nature *435*, 605–611.

Fenn, J.B., Mann, M., Meng, C.K., Wong, S.F., and Whitehouse, C.M. (1989). Electrospray ionization for mass spectrometry of large biomolecules. Science *246*, 64–71.

Garrisi, V.M., Abbate, I., Quaranta, M., Mangia, A., Tommasi, S., and Paradiso, A. (2008). SELDI-TOF serum proteomics and breast cancer: which perspective? Expert. Rev. Proteomics *5*, 779–785.
Goëb, V., Thomas-L'Otellier, M., Daveau, R., Charlionet, R., Fardellone, P., Le Loët, X., Tron, F., Gilbert, D., and Vittecoq, O. (2009). Candidate autoantigens identified by mass spectrometry in early rheumatoid arthritis are chaperones and citrullinated glycolytic enzymes. Arthritis Res. Ther. *11*, R38.
Gollapalli, K., Ray, S., Srivastava, R., Renu, D., Singh, P., Dhali, S., Bajpai, D.J., Srikanth, R., Moiyadi, A., and Srivastava, S. (2012). Investigation of serum proteome alterations in human glioblastoma multiforme. Proteomics. *12*, 2378–2390.
Graber, J.J., and Dhib-Jalbut, S. (2011). Biomarkers of disease activity in multiple sclerosis. J. Neurol. Sci. *305*, 1–10.
Gruys, E., Toussaint, M.J., Niewold, T.A., and Koopmans, S.J. (2005). Acute phase reaction and acute phase proteins. J. Zhejiang. Univ Sci. B *6*, 1045–1056.
Hall, D.A., Ptacek, J., and Snyder, M. (2007). Protein microarray technology. Mech. Ageing Dev. *128*, 161–167.
Harry, J.L., Wilkins, M.R., Herbert, B.R., Packer, N.H., Gooley, A.A., and Williams, K.L. (2000). Proteomics: capacity versus utility. Electrophoresis *21*, 1071–1081.
Hartmann, M., Roeraade, J., Stoll, D., Templin, M.F., and Joos, T.O. (2009). Protein microarrays for diagnostic assays. Anal. Bioanal. Chem. *393*, 1407–1416.
He, Q.Y., and Chiu, J.F. (2003). Proteomics in biomarker discovery and drug development. J. Cell Biochem. *89*, 868–886.
Hortin, G.L., Jortani, S.A., Ritchie, J.C., Jr., Valdes, R., Jr., and Chan, D.W. (2006). Proteomics: a new diagnostic frontier. Clin. Chem. *52*, 1218–1222.
Hueber, W., and Robinson, W.H. (2006). Proteomic biomarkers for autoimmune disease. Proteomics. *6*, 4100–4105.
Hutchens, T.W., and Yip, T.T. (1993). New desorption strategies for the mass spectrometric analysis of macromolecules. RapidCommun Mass Spectrom *7*, 576–580.
Issaq, H.J., Xiao, Z., and Veenstra, T.D. (2007). Serum and plasma proteomics. Chem. Rev. *107*, 3601–3620.
de Jager, W., and Rijkers, G.T. (2006). Solid-phase and bead-based cytokine immunoassay: a comparison. Methods *38*, 294–303.
James, P., Quadroni, M., Carafoli, E., and Gonnet, G. (1993). Protein identification by mass profile fingerprinting. Biochem. Biophys. Res. Commun. *195*, 58–64.
Jemal, A., Bray, F., Center, M.M., Ferlay, J., Ward, E., and Forman, D. (2011). Global cancer statistics. CA Cancer J. Clin. *61*, 69–90.
Kalinina, J., Peng, J., Ritchie, J.C., and Van Meir, E.G. (2011). Proteomics of gliomas: initial biomarker discovery and evolution of technology. Neuro. Oncol. *13*, 926–942.
Karas, M., and Hillenkamp, F. (1988). Laser desorption ionization of proteins with molecular masses exceeding 10,000 daltons. Anal. Chem. *60*, 2299–2301.
Kassa, F.A., Shio, M.T., Bellemare, M.J., Faye, B., Ndao, M., and Olivier, M. (2011). New inflammation-related biomarkers during malaria infection. PLoS. One. *6*, e26495.
Khan, S., Jutzy, J.M., Valenzuela, M.M., Turay, D., Aspe, J.R., Ashok, A., Mirshahidi, S., Mercola, D., Lilly, M.B., and Wall, N.R. (2012). Plasma-derived exosomal survivin, a plausible biomarker for early detection of prostate cancer. PLoS. One. *7*, e46737.
Klose, J. (1975). Protein mapping by combined isoelectric focusing and electrophoresis of mouse tissues. A novel approach to testing for induced point mutations in mammals. Humangenetik *26*, 231–243.
Kosako, H., Yamaguchi, N., Aranami, C., Ushiyama, M., Kose, S., Imamoto, N., Taniguchi, H., Nishida, E., and Hattori, S. (2009). Phosphoproteomics reveals new ERK MAP kinase targets and links ERK to nucleoporin-mediated nuclear transport. Nat. Struct. Mol. Biol. *16*, 1026–1035.
Kumar, D.M., Thota, B., Shinde, S.V., Prasanna, K.V., Hegde, A.S., Arivazhagan, A., Chandramouli, B.A., Santosh, V., and Somasundaram, K. (2010). Proteomic identification of haptoglobin alpha2 as a glioblastoma serum biomarker: implications in cancer cell migration and tumor growth. J. Proteome. Res. *9*, 5557–5567.
Laronga, C., Becker, S., Watson, P., Gregory, B., Cazares, L., Lynch, H., Perry, R.R., Wright, G.L., Jr., Drake, R.R., and Semmes, O.J. (2003). SELDI-TOF serum profiling for prognostic and diagnostic classification of breast cancers. Dis. Markers *19*, 229–238.
Lash, G.E., Scaife, P.J., Innes, B.A., Otun, H.A., Robson, S.C., Searle, R.F., and Bulmer, J.N. (2006). Comparison of three multiplex cytokine analysis systems: Luminex, SearchLight and FAST Quant. J. Immunol. Methods *309*, 205–208.

Lescuyer, P., Hochstrasser, D.F., and Sanchez, J.C. (2004). Comprehensive proteome analysis by chromatographic protein prefractionation. Electrophoresis *25*, 1125–1135.

van der Linden, M.P., van der Woude, D., Ioan-Facsinay, A., Levarht, E.W., Stoeken-Rijsbergen, G., Huizinga, T.W., Toes, R.E., and van der Helm-van Mil A.H. (2009). Value of anti-modified citrullinated vimentin and third-generation anti-cyclic citrullinated peptide compared with second-generation anti-cyclic citrullinated peptide and rheumatoid factor in predicting disease outcome in undifferentiated arthritis and rheumatoid arthritis. Arthritis Rheum. *60*, 2232–2241.

Luna Coronell, J.A., Syed, P., Sergelen, K., Gyurján, I., and Weinhäusel, A. (2012). The current status of cancer biomarker research using tumour-associated antigens for minimal invasive and early cancer diagnostics. J. Proteomics *76* Spec No., 102–115.

Mann, M., and Wilm, M. (1994). Error-tolerant identification of peptides in sequence databases by peptide sequence tags. Anal. Chem. *66*, 4390–4399.

Mann, M., Hojrup, P., and Roepstorff, P. (1993). Use of mass spectrometric molecular weight information to identify proteins in sequence databases. Biol. Mass Spectrom. *22*, 338–345.

Mayer, M.C., Hohlfeld, R., and Meinl, E. (2012). Viability of autoantibody-targets: how to tackle pathogenetic heterogeneity as an obstacle for treatment of multiple sclerosis. J. Neurol. Sci. *319*, 2–7.

Meinl, E., Krumbholz, M., and Hohlfeld, R. (2006). B lineage cells in the inflammatory central nervous system environment: migration, maintenance, local antibody production, and therapeutic modulation. Ann. Neurol. *59*, 880–892.

Nell, V.P., Machold, K.P., Stamm, T.A., Eberl, G., Heinzl, H., Uffmann, M., Smolen, J.S., and Steiner, G. (2005). Autoantibody profiling as early diagnostic and prognostic tool for rheumatoid arthritis. Ann. Rheum. Dis. *64*, 1731–1736.

Nesterova, M.V., Johnson, N., Cheadle, C., Bates, S.E., Mani, S., Stratakis, C.A., Khan, I.U., Gupta, R.K., and Cho-Chung, Y.S. (2006). Autoantibody cancer biomarker: extracellular protein kinase A. Cancer Res. *66*, 8971–8974.

O'Farrell, P.H. (1975). High resolution two-dimensional electrophoresis of proteins. J. Biol. Chem. *250*, 4007–4021.

Omenn, G.S. (2004). Advancement of biomarker discovery and validation through the HUPO plasma proteome project. Dis. Markers *20*, 131–134.

Opstal-van Winden, A.W., Krop, E.J., Karedal, M.H., Gast, M.C., Lindh, C.H., Jeppsson, M.C., Jonsson, B.A., Grobbee, D.E., Peeters, P.H., Beijnen, J.H., *et al.* (2011). Searching for early breast cancer biomarkers by serum protein profiling of pre-diagnostic serum; a nested case–control study. BMC. Cancer *11*, 381.

Ortea, I., Roschitzki, B., Ovalles, J.G., Longo, J.L., de la Torre, I., Gonzalez, I., Gomez-Reino, J.J., and Gonzalez, A. (2012). Discovery of serum proteomic biomarkers for prediction of response to infliximab (a monoclonal anti-TNF antibody) treatment in rheumatoid arthritis: an exploratory analysis. J. Proteomics *77*, 372–382.

Pappin, D.J., Hojrup, P., and Bleasby, A.J. (1993). Rapid identification of proteins by peptide-mass fingerprinting. Curr. Biol. *3*, 327–332.

Phizicky, E., Bastiaens, P.I., Zhu, H., Snyder, M., and Fields, S. (2003). Protein analysis on a proteomic scale. Nature *422*, 208–215.

Poruk, K.E., Firpo, M.A., Scaife, C.L., Adler, D.G., Emerson, L.L., Boucher, K.M., and Mulvihill, S.J. (2013). Serum osteopontin and tissue inhibitor of metalloproteinase 1 as diagnostic and prognostic biomarkers for pancreatic adenocarcinoma. Pancreas *42*, 193–197.

Rabilloud, T., Chevallet, M., Luche, S., and Lelong, C. (2010). Two-dimensional gel electrophoresis in proteomics: Past, present and future. J. Proteomics. *73*, 2064–2077.

Rai, A.J., Gelfand, C.A., Haywood, B.C., Warunek, D.J., Yi, J., Schuchard, M.D., Mehigh, R.J., Cockrill, S.L., Scott, G.B., Tammen, H., *et al.* (2005). HUPO Plasma Proteome Project specimen collection and handling: towards the standardization of parameters for plasma proteome samples. Proteomics *5*, 3262–3277.

Rath, A., Glibowicka, M., Nadeau, V.G., Chen, G., and Deber, C.M. (2009). Detergent binding explains anomalous SDS-PAGE migration of membrane proteins. Proc. Natl. Acad. Sci. U.S.A *106*, 1760–1765.

Ray, S., Reddy, P.J., Jain, R., Gollapalli, K., Moiyadi, A., and Srivastava, S. (2011). Proteomic technologies for the identification of disease biomarkers in serum: advances and challenges ahead. Proteomics. *11*, 2139–2161.

Ray, S., Kamath, K.S., Srivastava, R., Raghu, D., Gollapalli, K., Jain, R., Gupta, S.V., Ray, S., Taur, S., Dhali, S., *et al.* (2012a). Serum proteome analysis of vivax malaria: An insight into the disease pathogenesis and host immune response. J. Proteomics. *75*, 3063–3080.

Ray, S., Renu, D., Srivastava, R., Gollapalli, K., Taur, S., Jhaveri, T., Dhali, S., Chennareddy, S., Potla, A., Dikshit, J.B., *et al.* (2012b). Proteomic investigation of falciparum and vivax malaria for identification of surrogate protein markers. PLoS. One. *7*, e41751.

Ray, S., Srivastava, R., Tripathi, K., Vaibhav, V., Patankar, S., and Srivastava, S. (2012c). Serum proteome changes in dengue virus-infected patients from a dengue-endemic area of India: towards new molecular targets? OMICS. *16*, 527–536.

Robinson, W.H., Steinman, L., and Utz, P.J. (2002). Proteomics technologies for the study of autoimmune disease. Arthritis Rheum. *46*, 885–893.

Roepstorff, P. (2012). Mass spectrometry based proteomics, background, status and future needs. Protein Cell *3*, 641–647.

Sahin, U., Tureci, O., Schmitt, H., Cochlovius, B., Johannes, T., Schmits, R., Stenner, F., Luo, G., Schobert, I., and Pfreundschuh, M. (1995). Human neoplasms elicit multiple specific immune responses in the autologous host. Proc. Natl. Acad. Sci. U.S.A *92*, 11810–11813.

Sahin, U., Tureci, O., and Pfreundschuh, M. (1997). Serological identification of human tumor antigens. Curr. Opin. Immunol. *9*, 709–716.

Sanchez-Carbayo, M. (2006). Antibody arrays: technical considerations and clinical applications in cancer. Clin. Chem. *52*, 1651–1659.

Sanchez-Carbayo, M. (2010). Antibody array-based technologies for cancer protein profiling and functional proteomic analyses using serum and tissue specimens. Tumour. Biol. *31*, 103–112.

Schett, G., and Gravallese, E. (2012). Bone erosion in rheumatoid arthritis: mechanisms, diagnosis and treatment. Nat. Rev. Rheumatol. *8*, 656–664.

Schiffer, E., Mischak, H., and Novak, J. (2006). High resolution proteome/peptidome analysis of body fluids by capillary electrophoresis coupled with MS. Proteomics. *6*, 5615–5627.

Scott, D.L., Wolfe, F., and Huizinga, T.W. (2010). Rheumatoid arthritis. Lancet *376*, 1094–1108.

Seurynck-Servoss, S.L., White, A.M., Baird, C.L., Rodland, K.D., and Zangar, R.C. (2007). Evaluation of surface chemistries for antibody microarrays. Anal. Biochem. *371*, 105–115.

Soydinc, H.O., Duranyildiz, D., Guney, N., Derin, D., and Yasasever, V. (2012). Utility of serum and urine uPAR levels for diagnosis of breast cancer. Asian Pac. J. Cancer Prev. *13*, 2887–2889.

Srivastava, R., Ray, S., Vaibhav, V., Gollapalli, K., Jhaveri, T., Taur, S., Dhali, S., Gogtay, N., Thatte, U., Srikanth, R., and Srivastava, S. (2012). Serum profiling of leptospirosis patients to investigate proteomic alterations. J. Proteomics *76* Spec No., 56–68.

Stempfer, R., Syed, P., Vierlinger, K., Pichler, R., Meese, E., Leidinger, P., Ludwig, N., Kriegner, A., Nöhammer, C., and Weinhäusel, A. (2010). Tumour auto-antibody screening: performance of protein microarrays using SEREX derived antigens. BMC. Cancer *10*, 627.

Sun, J.F., Shi, Z.X., Guo, H.C., Li, S., and Tu, C.C. (2011). Proteomic analysis of swine serum following highly virulent classical swine fever virus infection. Virol. J. *8*, 107.

Syed, P. (2012). Serum auto-antibody testing for early diagnosis of breast cancer. PhD Thesis, Vienna University of Technology, Faculty of Technical Chemistry.

Szodoray, P., and Alex, P. (2011). Protein array diagnostics for guiding therapy in rheumatoid arthritis. Mol. Diagn. Ther. *15*, 247–254.

Tan, H.T., Low, J., Lim, S.G., and Chung, M.C. (2009). Serum autoantibodies as biomarkers for early cancer detection. FEBS J. *276*, 6880–6904.

Tan, H.T., Lee, Y.H., and Chung, M.C. (2012). Cancer proteomics. Mass Spectrom. Rev. *31*, 583–605.

Tanaka, T., Sakurada, S., Kano, K., Takahashi, E., Yasuda, K., Hirano, H., Kaburagi, Y., Kobayashi, N., Hang, N.T., Lien, L.T., *et al.* (2011). Identification of tuberculosis-associated proteins in whole blood supernatant. BMC. Infect. Dis. *11*, 71.

TDR Diagnostics Evaluation Expert Panel, Banoo, S., Bell, D., Bossuyt, P., Herring, A., Mabey, D., Poole, F., Smith, P.G., Sriram, N., Wongsrichanalai, C., *et al.* (2010). Evaluation of diagnostic tests for infectious diseases: general principles. Nat. Rev. Microbiol. *8*, S17-S29.

Unlü, M., Morgan, M.E., and Minden, J.S. (1997). Difference gel electrophoresis: a single gel method for detecting changes in protein extracts. Electrophoresis *18*, 2071–2077.

Vogel, C., Abreu, R.S., Ko, D., Le, S.Y., Shapiro, B.A., Burns, S.C., Sandhu, D., Boutz, D.R., Marcotte, E.M., and Penalva, L.O. (2010). Sequence signatures and mRNA concentration can explain two-thirds of protein abundance variation in a human cell line. Mol. Syst. Biol. *6*, 400.

Wandall, H.H., Blixt, O., Tarp, M.A., Pedersen, J.W., Bennett, E.P., Mandel, U., Ragupathi, G., Livingston, P.O., Hollingsworth, M.A., Taylor-Papadimitriou, J., *et al.* (2010). Cancer biomarkers defined by autoantibody signatures to aberrant O-glycopeptide epitopes. Cancer Res. *70*, 1306–1313.

Waterboer, T., Sehr, P., and Pawlita, M. (2006). Suppression of non-specific binding in serological Luminex assays. J. Immunol. Methods *309*, 200–204.

Yates, J.R., Ruse, C.I., and Nakorchevsky, A. (2009). Proteomics by mass spectrometry: approaches, advances, and applications. Annu. Rev. Biomed. Eng *11*, 49–79.

Zhang, A., Sun, H., Wang, P., and Wang, X. (2013). Salivary proteomics in biomedical research. Clin. Chim. Acta *415*, 261–265.

Zhang, H., Liu, A.Y., Loriaux, P., Wollscheid, B., Zhou, Y., Watts, J.D., and Aebersold, R. (2007). Mass spectrometric detection of tissue proteins in plasma. Mol. Cell Proteomics. *6*, 64–71.

Zhu, H., and Snyder, M. (2003). Protein chip technology. Curr. Opin. Chem. Biol. *7*, 55–63.

Serum Profiling by Targeted Proteomics for Biomarker Discovery

2

Paula Díez, Maria Gonzalez-Gonzalez, Noelia Dasilva, Ricardo Jara-Acevedo, Alberto Orfao and Manuel Fuentes

Abstract

Personalized medicine is one of the most important goals in medical research. Advances in detection and treatment of diseases have improved during the last decades, but there are still difficulties regarding the response to therapies depending on individual profiles. With the purpose of elucidating the best methods to deal with pathologies, various disciplines have emerged. Proteomics appears as the most promising in biomarker discovery. Detecting proteins with biomarker utility may be essential for a better diagnosis and prognosis. In this review we describe several targeted proteomic quantification methodologies. In addition, we summarize studies in cancer, autoimmune and metabolic diseases in which mass spectrometry approaches have been employed for biomarker discovery.

Biomarker discovery

Having deciphered the human genome, determining the proteins encoded by genes is the principal purpose established in scientific disciplines. Proteomics focuses on the characterization of thousands of proteins in biological samples. And, specifically, these efforts aim to deal with the number of proteins within a complex proteome, which is unpredictable. It is not only the huge number of proteins, but also the need for specific and sensitive identification methods, that makes it difficult to study the necessary targeted proteomics.

Depending on the kind of sample, protein analysis gets difficult or easier according to the complexity, sample extraction method, sample viability, lack of standard manufacturing steps, protein modifications, relative abundance, and validation methods.

For example, blood biomarkers are characterized by their dynamic range of concentrations present in human plasma, what difficult their study. The complexity of this kind of samples makes the analysis more complicated, too. A great amount of different proteins are identified at once, so verification techniques are required. Thousands of proteins are identified but just a few of them could be established as biomarkers. Furthermore, if these biomarkers are used in clinical routine, the number of verifications is increased in a proportional way with the number of patient samples. The most commonly technique used with this purpose is enzyme-linked immunosorbent assay (ELISA). Its sensitivity and specificity justify its clinical use, but the drawbacks are remarkable: its high cost and time-consuming make necessary new detection strategies (Huttenhain *et al.*, 2009).

It is not only traditional techniques that present disadvantages; new methods, such as high-performance liquid chromatography (HPLC) coupled to a mass spectrometry (MS) in a data-dependent acquisition (DDA), also have their drawbacks, the principal ones being poor reproducibility and failure to identify low-abundance proteins (Nahnsen and Kohlbacher, 2012).

Proteomics offers different MS-based strategies for biomarker identification (Fig. 2.1). These are now briefly summarized (Veenstra, 2007):

1 *Two-dimensional PAGE coupled to MS.* Protein samples are separated according to their isoelectric point and molecular mass using 2D-PAGE. After in-gel tryptic digestion, protein spots are analysed by MS approaches. This classical coupled technique still presents limitations in sensitivity. Lower abundance proteins are also masked, preventing their detection. Therefore, enrichment of these low abundance proteins is a requirement prior to 2D-PAGE.
2 *Proteomic profiling.* In this strategy, proteins are bound directly to a chip and their mass spectrum is obtained with a time-of-flight (TOF) spectrometer. An algorithm compares the results from the sample of interest and the control sample giving information about the abundance of a specific protein. This technique is mainly characterized by its high throughput (HT), but the detection of low-abundance proteins is limited by the presence of high-abundance proteins in the sample.
3 *Stable-isotope proteome tagging.* In isotope-coded affinity tags (ICATs) proteomes from two different samples are labelled with reagents that are chemically and functionally the same, but differ in mass (heavy and light ICAT reagents). After digestion, stable-isotope

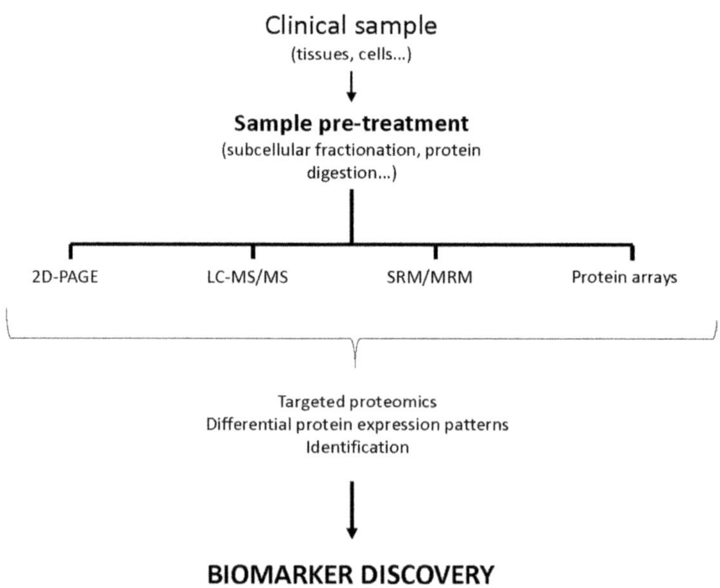

Figure 2.1 Pipeline of biomarker discovery. Clinical samples are pre-treated and analysed by using different proteomics tools (2D-PAGE, LC-MS/MS, SRM/MRM, protein array, etc.) to discover disease biomarkers which could lead to personalized medicine.

tagged peptides are extracted with an avidin column. Biotin is removed and peptides are analysed by multidimensional chromatography coupled to MS/MS. The principal drawback is the masking of low abundance proteins, which are not labelled with ICAT reagents.

4 *Subtractive proteomics*. The basis of this strategy is the relation between the numbers of peptides from each sample. After protein extraction and tryptic digestion, peptides from each protein obtained in the sample of interest are compared to those peptides obtained in the healthy control sample. The resulting relation determines the abundance of the protein in the sample. The main disadvantage is related to low-abundance proteins, it is necessary, at least, 3-fold changes to determines, with high confidence, the relative abundance for a given protein. Furthermore, it is a time-consuming technique.

In summary, the main drawback common to all these strategies is the masking of low-abundance proteins. Therefore, removing high-abundance proteins could reduce the complexity of the sample improving its analysis. Albumin and immunoglobulins have been identified as the most abundant proteins in plasma by the HUPO plasma proteome initiative. Selective immunodepletion has been used by Brand *et al.* (2006) using a column format providing a 5-fold enrichment of lower abundance proteins by removing 85% of the total plasma protein. Further to these results, Sigma-Aldrich has developed an immunodepletion column capable of removing the 20 most abundant proteins. These improvements have allowed advances in biomarker discovery, but there are still limitations in the low ng/ml concentration range (Schiess *et al.*, 2009).

Identifying thousands of proteins in a sample is not enough for biomarker detection. Validation of biomarker candidates is one of the most critical points in biomarker prediction. Immuno- and non-immuno-based methods allow the confirmation of the biomarker nature of a protein (Poschmann *et al.*, 2009).

- *Immuno-based strategies*. ELISA is the most popular and sensitive (20 pg/ml) method for validation, although it is still costly and time-consuming. It is based on a union between an antibody and the protein of interest, amplifying the signal. Western blotting works in a similar way. It is capable of giving information about protein modifications. Poschmann *et al.* (2009) prefer immunohistochemistry (IHC) on tissue microarray (TMA) owing to its possibility of conducting analyses in parallel, favouring comparisons.
- *Non-immuno-based strategies*. Validating biomarker candidates without using antibodies is other option, and MS-based multi-reaction monitoring (MRM) has become the method of election. It will be described in detail in the next section.

All in all, discovering diagnostic and prognostic biomarkers may provide useful information in clinical area improving the correct election of a therapy which could make conditional to patient's survival. The search for personalized medicine continues being the main goal, and therefore, all efforts are focused on the development of strategies and methodologies which could identify proteins in a rapid, easy, non-time-consuming and reliable fashion.

In order to introduce the next sections, we define here the protein biomarker as the protein present in fluids or tissues which is expressed under determined conditions reflecting the presence of pathology. The concept can be used in its positive form too. Nevertheless, we will employ 'protein biomarker' for proteins reflecting disease.

Kulasingam and Diamandis (2008) classify protein biomarker into three principal categories:

1 *Diagnostic screening biomarkers:* proteins for cancer detection in an individual with high sensitivity and specificity.
2 *Prognostic biomarkers:* once diagnosis has been made, these biomarkers give information about the progression of the disease, helping in the treatment election.
3 *Stratification biomarkers:* they predict response to therapy, dividing patients in groups in function of their response.

Proteomics for diagnostic biomarkers

Medical diagnosis is the first step at patient's classification that must be followed to establish the treatment and prognosis conditions. Depending on the disease, different techniques are developed to detect and confirm the pathology. As soon as a diagnosis is made, a therapy can be applied, improving the prognosis and the quality of life of patients.

Cancer diagnosis is one of the principal areas in which proteomics are focused on. The principal drawbacks lie in the heterogeneous nature of the disease: patient variability, complexity of the samples, absence of technologies for sensitive and specific biomarker discovery, and lack of optimized validation methods (Pin *et al.*, 2013). Detection in early stages would provide a better prognosis, for this reason all efforts in this field are focused on the detection of tumour biomarkers which can lead to the best treatment.

E. Pin and collaborators (Pin *et al.*, 2013) have reviewed the recent advances in prostate cancer (PCa) research. Further than prostate-specific antigen (PSA), it is necessary to discover new biomarkers that allow distinguish between aggressive and indolent cancers to decide which the best treatment is. It is known that standard treatment involves several therapies (radical prostatectomy, hormonal and radiation therapy, chemotherapy or cryosurgery) that are not necessary in all cases. Its application is conditioned by the lack of awareness about the real cause of the cancer. Discovering other biomarkers could allow the determination of the cancer stage, the origin of the disease and the election of the best treatment for each patient, or what is the same biomarker discovery would lead to personalized medicine.

By using MS methodologies, such as 2DE-MS, matrix-associated laser desorption ionization (MALDI)-MS and surface-enhanced laser desorption/ionization (SELDI)-MS, several new PCa biomarkers have been detected. E. Pin summarizes in his review the principal biological fluids and tissue biomarkers for PCa, among which are calgranulin, uromodulin, afamin or CD90 for biological fluids, and prostate-specific membrane antigen (PSMA), GDF15 or PCa24 for tissues.

Somparn *et al.* (2012) have applied proteomics to urinary system for the diagnosis of active lupus nephritis, also known as systemic lupus erythematosus (SLE). 2-DE analysis was performed in urine samples for the initial screening. They analysed the differential protein expression between active and inactive patients and the selected spots were identified by electrospray ionization (ESI)-quadrupole (Q)- TOF MS/MS. Finally, they detected prostaglandin H_2D-isomerase (PGDS) and three isoforms of Zn-α2-glycoprotein (ZA2G) increased in the active SLE condition. After validation, PGDS was established as a possible specific biomarker for the disease.

In contrast, Blanco's lab (Ruiz-Romero and Blanco, 2010) screens plasma, serum and synovial fluid searching for biomarkers of osteoarthritis (OA). They have identified, for example, structural proteins of connective tissues that could determine the initial development of the OA improving the diagnosis and, therefore, the prognosis.

In Niemann–Pick type C (NPC) disease lysosomal storage is affected as a result of a mutation in *NPC1* gene. The diagnosis used to be made by invasive approaches. Maekawa *et al.* (2013) have identified a biomarker in urine by LC/ESI-MS/MS. This biomarker, 3β-sulfooxy-7β-*N*-acetylglucosaminyl-5-cholen-24-oic acid, and its conjugates, occur in high concentrations in NPC patients.

Proteomics for prognostic biomarkers

Once a disease has been characterized and diagnosed in an individual, prognosis is the next step, which deals with the prediction of probable outcomes of the medical situation. Depending on the outcomes, treatment could change. For this reason, the need for new approaches to biomarker discovery is once again evident.

Kikuta and colleagues (Kikuta *et al.*, 2012) have focused their work on gastrointestinal stromal tumour (GIST). This kind of cancer is the most common mesenchymal malignancy of the gastrointestinal tract. Surgery is the usual approach of treatment in this pathology. They have studied the development of metastasis after surgery, looking for proteomics characteristics which could become prognosis biomarkers. Their discoveries have revealed the utility of ATP-dependent RNA helicase DDX39 as biomarker for metastasis and poor clinical outcomes in GIST patients.

Cholangiocarcinoma is another kind of cancer that occurs worldwide. The more effective treatment is gemcitabine, but it is beneficial only for a limited subset of patients. For this reason, the proteome characterization of patients is undoubtedly necessary to discover biomarkers for predicting the response to gemcitabine. MS was the strategy chosen by Morofuji *et al.* (2012) to determine this relation between the drug and the prognosis. In summary, they have determined the use of CapG as a predictive biomarker in prognosis. Patients who are CapG-positive require an increase in their gemcitabine therapy.

A similar study has been developed by Cheng and collaborators (Cheng *et al.*, 2011) in hepatocellular carcinoma. Their proteomic analysis have revealed the correlation between the protein codified by N-myc downstream-regulated gene 1 (NDRG1), metastasis development and poor prognosis of patients after liver transplantation.

Identifying drug resistances is other important task in prognosis. Breast, lung and ovarian cancers are several examples in which biomarker discovery studies have been done.

Protein profiles were studied before and after paclitaxel was administered to breast cancer patients in order to analyse the differences between protein expression patterns. Patients with a complete response to the treatment show high levels of alpha-defensins (DEFA) and microtubule-associated protein 2 (MAP2). Therefore, these proteins could be used as putative predictive biomarkers for response to chemotherapy (Bauer *et al.*, 2010). Other studies have been performed in this cancer for prediction of response to treatment, highlighting studies in tamoxifen resistance.

As in breast cancer, paclitaxel is the treatment of choice in ovarian cancer. Studies have demonstrated the differential expression of tumour rejection antigen 1, triose phosphate isomerase, palmitoyl-protein thioesterase 1 precursor and ER-associated DNAJ and their utility as predictive biomarkers (Di Michele *et al.*, 2010).

One of the more effective treatments in lung cancer is gemcitabine-based chemotherapy. In spite of that, many patients cannot benefit from it. Proteomic studies have showed the association between sorcin overexpression and gemcitabine-resistance and poor prognosis in lung cancer patients. Cisplatin resistance was also identified, revealing DJ-1 as a possible predictive biomarker (Li *et al.*, 2011).

Proteomics for personalized medicine

Personalized medicine is defined as the group of medical decisions for the benefit of the individual profile. Medical monitoring and treatment are tailored according to patient features and pathology.

Variations in response to treatments, even at the same stage of the disease, are due to intrinsic differences between individuals. The knowledge about these fluctuations may manage to a better prognosis.

This field is still in its infancy, but proteomics appears to be the most promising technology to achieve the objective. The first step consists in the establishment of subgroups for disease. With this preliminary classification, patients could be included in a defined set and then, taking into account other medical considerations, therapy and treatment could be selected according to the specific patient's situation.

Cancer, obesity and diabetes, transplant rejection and hypertension are several examples of diseases that show large differences in complexity between individuals despite similar clinical features (Nicolini *et al.*, 2012).

Targeted proteomics

Proteomic approaches have been established as powerful tools for systems biology and biomarker discovery. However, there are several limitations related to quantification accuracy and the large number of data generated by these methodologies (Shi *et al.*, 2012). Targeted proteomics, in which a specific peptide is selected by its mass/charge (m/z) ratio, seems to be the most suitable methodology for protein quantification. In addition, results obtained with proteomic tools present limited statistical confidence. In spite of that, a large quantity of potential biomarkers has been described during the last years. The challenge appears when verification and validation techniques try to be applied. Additionally, low abundance, post-translational modifications (PTMs) and the lack of suitable affinity detection reagents constitute several drawbacks in quantitative measurement of proteins in signalling networks (Zhao and Brasier, 2013).

All in all, new strategies are required in this field. Targeted proteomics is the latest, selected reaction monitoring (SRM) being the most common option. SRM can quantify a target peptide with high confidence even in a complex sample. Its principal advantages are the sensitivity, dynamic range and reproducibility, which make it a promising tool for biomarker validation in blood plasma (Anderson and Hunter, 2006).

SRM and MRM assays

SRM technology

SRM is a MS technique for the targeted detection and accurate quantification of selected proteotypic peptides (PTPs) with known fragmentation properties in a complex sample matrix (Fig. 2.2). PTPs are unique for a given protein with respect to a whole proteome. First of all, proteins must be enzymatically digested in order to be separated by using liquid

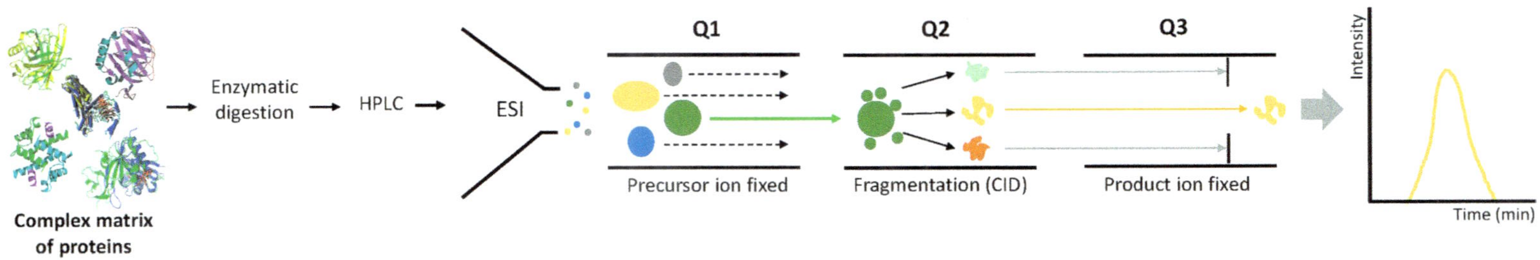

Figure 2.2 Selected reaction monitoring (SRM) and multiple reaction monitoring (MRM). A complex matrix is ionized in a electrospray ionization (ESI) and introduced in the triple quadrupole (QQQ). The first quadrupole (Q1) select the precursor ion and transmit it to the second quadrupole (Q2), in which the specific collision energy produces the protein fragmentation into peptides that are analysed in the mass analyser (Q3). Mass spectrum of the product ion is analysed. MRM works in the same way, but analysing many proteins at the same time.

chromatography and monitored with a triple quadrupole (Huttenhain *et al.*, 2009; Nahnsen and Kohlbacher, 2012).

Transitions are the technical basis of this strategy. Transitions are constituted by pairs of precursor/fragment ion of a determined *m/z* ratio. They are measured on a triple-quadrupole mass spectrometer. In the first quadrupole, precursor ions are selectively chosen; the fragmentation via collisional induced disassociation (CID) takes place in the second quadrupole, and the monitoring of specific fragments of ions occurs in the third one, the mass analyser quadrupole (Anderson and Hunter, 2006). In summary, a single product ion derived from the MS/MS fragmentation of the parent ion is measured in a SRM analysis.

For several decades, SRM has been utilized in disciplines of toxicology and pharmacokinetics (Yocum and Chinnaiyan, 2009). It is only within the last few years that it has been applied in quantitative proteomics area with a high selectivity, low background signals and high duty cycle. The impact of this technique in the mentioned area resides in its capability of identifying proteins in a wide dynamic range with a high degree of selectivity. In addition, the duty cycle allows, in multiplexed analyses, the acquisition of analytes that co-elute from the HPLC column, although their concentration is significantly different (Holman *et al.*, 2012).

Unlike Western blotting techniques, SRM assays do not require high-affinity antibodies and are capable of detecting protein concentrations in the pg/ml range. Moreover, the noise level is reduced thanks to pre-selection of precursor–product ion transitions (Zhao and Brasier, 2013). Due to high sensitivity and specificity, simplicity of assay configuration, and exceptional multiplexing capability, SRM has become the most frequent strategy used for low abundance biomarker discovery (Liu *et al.*, 2012).

Combinations of protein and peptide enrichment strategies with SRM have allowed the improvement of the technique. In this way, the complexity of the sample is reduced, but the capability of HT analysis is not altered. Several combined strategies are briefly summarized here (Table 2.1):

- *Immunoaffinity-SRM approach.* Peptide immunoaffinity enrichment coupled to SRM has emerged as a promising technology for protein quantification in the low ng/ml range (Whiteaker *et al.*, 2012). Between the possible strategies in this field, Nicol *et al.* (2008) have developed hydrazide beads that are used as substrate for the immobilization of specific antibodies. On the other side, Rafalko and collaborators have applied an enrichment step prior to LC-SRM based on magnetic bead-based antibodies which offer more sensitivity and specificity (Rafalko *et al.*, 2010).

Table 2.1 List of different strategies coupled to SRM

Couple-SRM technique	Advantages
Immunoaffinity-SRM approach	Protein quantification in the low ng/ml range
Stable isotope standards and capture by anti-peptide antibodies (SISCAPA)	High precision and accuracy, less cost, no depletion
N-glycosite subproteome	Glycoprotein quantification in the ng/ml range
Stable isotope dilution (SID)-SRM-MS	Accurate quantification of target proteins in a complex background

- *Stable isotope standards and capture by anti-peptide antibodies (SISCAPA).* The affinity enrichment of peptides is made with stable isotope dilution and detected by SRM (Gillette and Carr, 2013). Anti-peptides antibodies are employed to enrich the sample (Zhao *et al.*, 2011). Gillette and Carr (2013) were the pioneers in this technique. They captured proteins of interest and eluted them by using nanoaffinity columns covered with anti-peptides antibodies. Once the sample has been enriched (10^2- to 10^3-fold enrichment), SRM analysis is performed to quantify the enriched analyte by using a spiked-in isotopically labelled synthetic peptide. Later, Whiteaker *et al.* (2011) optimized SISCAPA in order to gain in precision and accuracy adapting the technique to a magnetic bead format (Schoenherr *et al.*, 2010; Whiteaker *et al.*, 2011; Yocum and Chinnaiyan, 2009). The main drawback of this modification is the requirement of one antibody for each peptide of interest. Indeed, monoclonal antibodies constitute a better alternative than polyclonal antibodies (Anderson *et al.*, 2004). But the advantages are remarkable: less cost, high analyte specificity and sensitivity, multiplexed assays, depletion is not required and a wide number of proteins and modifications can be analysed (Zhao *et al.*, 2011).
- *N-glycosite subproteome.* This is based in the fact that the great majority of proteins are glycosylated. The quantification of the glycoproteins is in the 5–100 ng/ml range (Stahl-Zeng *et al.*, 2007).
- *Stable isotope dilution (SID)-SRM-MS.* Its main application is quantifying low abundance proteins in signalling pathways. This combination is considered the most accurate approach for quantifying, in a direct way, target proteins in a complex background. A specific application of SID-SRM-MS assays is the measurement of activating PTMs. However, PTM profiling is not an easy task. Lack of ideal signature peptides, presence of reactive residues or more than one basic amino acid residue, and poor sensitivity are various of the problems of applying SID-SRM-MS to PTMs studies. The solution is based on the selection of signature peptides and optimization of their transitions (Zhao and Brasier, 2013).

Prior to analysis, it is necessary the definition of the peptide of interest. Several factors must be taken into account: (i) the peptide must be unique for the analysed proteome; (ii) the sample should have no contaminant proteins, as keratin; and (iii) quantitative and qualitative information required about the peptides must be established.

Once the definition of the peptides has been made, the next step is the design of the SRM experiment (Holman *et al.*, 2012). Four pillars are essential:

- The state of the precursor ion-charge for a given peptide.
- The *m/z* ratio of the ion generated from the precursor by low-energy CID. SRM technology looks for specific peptides as a function of their *m/z* ratio, which is characteristic for each peptide.
- The collision energy required to dissociate the precursor and generate optimal signal intensity of the product ion. Proteins can be dissociated according to its collision energy, which is the responsible of fragmenting the protein and generating the peptide fragments. Depending on the collision energy, proteins can be dissociated or remain stable.
- Retention time of the peptide for the identification of the correct peak in the MS-spectrum.

The development of SRM assays has allowed the analysis, detection and screening of signature peptides in relevant diseases. Thus, by using a triple quadrupole coupled to SRM, Heywood and collaborators have detected an elevated expression of serum amyloid-P (SAP) and plasma C1-inhibitor protein in the maternal plasma of pregnant women whose Down syndrome test was positive (Heywood *et al.*, 2012). In a similar way, Kroksveen *et al.* (2013) have analysed cerebrospinal fluid samples from patients in an early multiple sclerosis stage. They have dealt with identifying proteins with differential expression pattern in comparison with controls founding 11 proteins, three of them highly significant identified by SRM (alpha-1-antichymotrypsin, contactin-1, and apolipoprotein D).

Androgen-driven prostate cancer has been studied by using an SRM strategy (SID-LC/ESI/SRM/MS). The main purpose has been the quantification of androgen metabolites and their derivatives. Quantifying androgen metabolome and identifying possible biomarkers may lead to the development of new strategies in this kind of cancer, avoiding orchiectomy or chemical castration as solutions (Tamae *et al.*, 2013). For PSA identification without using specific antibodies, Liu *et al.* (2012) have developed, for the first time, an immunoaffinity depletion-SRM-MS assay. They have detected PSA in ng/ml range, validating the results with conventional immunoassays.

MRM technology

Multiple reaction monitoring (MRM) is a variant of SRM. MRM-MS while an older methodology is currently being employed in proteomics area. Traditionally, it has been utilized for quantitative analysis of small molecules. At present, it use is focused on protein identification and quantification. In contrast to SRM, MRM is based on monitoring multiple transitions within one MS run. The selection of MRM transitions is an important step, not all the fragment ions are always selected. For that reason, selected transition ions should maximize specificity and minimize interferences from co-eluting species in the same mass windows and tolerances. There are several recommendations for the design of MRM transitions:

1 Doubly charged parent ions for ESI. If there is a histidine residue in the peptide sequence or the mass is greater than 1500 Da, triply charged is required.
2 y-ions over b-ions.
3 Q3 fragment ion ought to have a greater *m/z* than the selected Q1 parent ion and as least four amino acids.
4 Three transitions per peptide, three peptides per protein having at least one unique sequence peptide.
5 Generally, one transition is used for quantification, whereas two transitions allow the identification of the peptide.

Related to quantification, MRM-MS is based on the classic isotope-dilution MS. Prior to o during the proteolytic step, labelled internal standard peptides must be introduced to the protein sample. Then, both samples, the labelled standard and the unlabelled protein sample, are analysed by LC-MRM-MS in a similar way to small molecule isotope dilution strategies. The absolute amount of the peptide of interest is easily deduced by calculating the ratio between the internal standard, of which concentration is known, and the native peptide. Additionally, precise quantification of protein isoforms is obtained with MRM methodology, concluding in a more sensitive, specific and selective information.

Despite the advantages of MRM-MS, it has several drawbacks. The time needed for method development, optimization and validation for quantification of the protein markers are the main ones. For this purpose, a novel hybrid instrument allows time reduction. It integrates a triple quadrupole and a linear trap (Le Blanc *et al.*, 2003).

An enormous challenge in proteomics has always been the identification of PTMs. MRM-MS allows the study of phosphorylation, acetylation and glycosylation. In this last case, for example, Stahl-Zeng *et al.* (2007) employed a combination of *N*-glycoside isolation and stable-isotope-labelled standard MRM-MS in order to identify glycosylations.

SRM versus MRM

With all data described above, we can summarize that fundamentally SRM and MRM are similar. The principal difference between these two targeted strategies is the number of measured product ions. In SRM only a single product ion derived from the MS/MS fragmentation of the parent ion is measured, whereas for MRM, many product ions are simultaneously measured.

SRM and MRM are based on the known of the *m/z* ratio, which allows the detection of the protein of interest improving the biomarker-discovery validation (Ye *et al.*, 2009).

Serum profiling by targeted proteomics

Serum profiling is based in searching protein expression patterns. Serum is the most common sample used in routine medical procedures thanks to its low invasive accession. Having in mind the serum complexity, it is the physiological fluidics which interplays with tissues and organs reflecting all alterations that could occur in those. All in all, serum has become a good alternative to invasive methodologies for studying cancer, autoimmune and metabolic diseases. The complexity of serum makes necessary MS-based methods, just like powerful data analysis software.

Several studies have been performed to establish efficient workflows for the profiling, purification and identification of proteins in human sera.

Cancer diseases

Schwamborn and coworkers were the first to successfully classify bladder carcinoma by analysing serum profiles by MS techniques. Specifically, serum pre-fractionation by MALDI-TOF MS was employed as a screening tool (Schwamborn *et al.*, 2009).

Li *et al.* (2012) have detected isoforms of serum amyloid protein A with high resolution and accuracy, all of them involved in several kinds of cancer including nasopharyngeal cancer, ovarian cancer, lung cancer, pancreatic cancer, prostate cancer, colorectal cancer, breast cancer, gastric cancer, paediatric osteosarcoma and neuroblastoma.

ESI-MS was used in another study to establish the differences between the sera of lung cancer patients and healthy control subjects. Investigators present the advantages of ESI-MS over SELDI/MALDI-MS, the traditional method for serum profiling. ESI-MS requires less sample manipulation, does not present random crystallization and requires no chemical additives (Hocker *et al.*, 2011).

SELDI-TOF-MS was employed for serum screening in neuroblastoma, comparing serum profiles from patients and healthy donors. Sandoval and collaborators show in their research article the MS spectra pointing out the peaks with differential expression patterns between mentioned conditions. They proposed their method as a robust tool for differentiation (Sandoval *et al.*, 2006).

In a previous review published by our group (Dasilva *et al.*, 2012), we reported on several tumour biomarkers detected by targeted proteomics, for example hnRNPK and PTHA proteins for head and neck cancer, CA-125 for ovarian cancer, and urinary calgranulin B/MRP-14 for prostate cancer.

Autoimmune diseases

Autoimmune diseases arise from an inappropriate immune response against own substances and tissues normally present in the body. An autoimmune disease must meet Witebsky's postulates (Witebsky *et al.*, 1957):

- direct evidence of transfer of pathogenic antibody or pathogenic T-cells;
- indirect evidence based on reproduction of the autoimmune disease in experimental animals;
- circumstantial evidence from clinical clues.

There are more than 80 autoimmune diseases and they are among the 10 leading causes of death. One of the most common autoimmune diseases is OA, a chronic degenerative joint disease which causes pain, stiffness, crepitus, swelling and disability. Articular cartilage suffers gradual degradation principally due to thymosine β4. This protein is involved in actin polymerization, wound healing, apoptosis, inflammatory responses and tumour metastasis, among others. Its function in cartilage area is the activation of matrix metalloproteinase-2 (MMP-2) and MMP-9, leading to cartilage destruction. Wei and co-workers have confirmed the correlation between thymosine β4 serum expression level and disease severity observed by radiography (Wei *et al.*, 2013).

Synovial fluid, which is a serum filtrate located in the joint, from OA patients has been studied by Mateos *et al.* (2012) to determine differential protein profiles. Proteins related to the formation and remodelling of the extracellular matrix (i.e. fibronectin, kininogen-1) were found increased in OA. They set these proteins as putative disease biomarkers (Mateos *et al.*, 2012).

Atopic dermatitis is a chronic inflammatory skin disease characterized by pruritus and eczemas. Infiltration of eosinophils, macrophages and T lymphocytes, and high levels of soluble IL-2 receptor and thymus activation-regulated chemokine, among others, have been reported to be correlated with the disease (Kakinuma *et al.*, 2003). Serum vitamin D levels seem to be correlated with the severity of atopic dermatitis. Lee *et al.* (2013) performed proteomic approaches; specifically LC coupled to MS/MS, and found the mentioned correlation with food sensitization.

As reported previously in our review (Dasilva *et al.*, 2012), α-enolase, haptoglobin and serum amyloid A are associated with Behçet's disease, transferrin, ceruloplasmin and GAPDH with juvenile idiopathic arthritis, and peptides of C-reactive protein with rheumatoid disease as putative biomarkers that can detected by targeted proteomic approaches.

Metabolic diseases

Diabetes is a metabolic disease characterized by hyperglycaemia. There are several types of diabetes. Herein, we mention type 2 diabetes mellitus (T2DM) and the study performed by Kaur and colleagues. By using a combination of isobaric tag for relative and absolute quantitation (iTRAQ) shotgun and MRM-based targeted quantitative proteomics, they have characterized various T2DM biomarkers. Vitamin D-binding protein was found to be highly expressed in sera from patients with diabetes. In addition, up-regulation of transthyretin and retinol-binding protein 4 was reported (Kaur *et al.*, 2012).

Conclusions

Biomarker discovery is one of the principal challenges in personalized medicine. As it has been shown over the chapter, proteomics is the most promising methodology with this purpose and, specifically, targeted proteomics.

On the one hand, SRM and MRM techniques have been established as the best tools for quantification in the ng/ml range, allowing the detection of low abundance proteins which could be work as biomarkers.

On the other hand, serum has been demonstrated to be the most suitable kind of sample for biomarker discovery. Its easy accessibility and its capability of presenting the environment molecular variations make serum a good choice.

Many diseases, such as cancer, autoimmune diseases and metabolic diseases, have been studied by applying targeted proteomics to serum profiling. In this way, many biomarkers for diagnosis, prognosis and choice of treatment have been discovered.

In the present era, in which advances in medical research are evident, society requires personalized medicine. The possibility of applying specific therapies depending on the individual conditions might lead to a better medicine. Emphasis is made in diagnosis due to its importance in the treatment of the diseases. In the most cases, the earlier a disease is diagnosed, the better is the prognosis.

References

Anderson, L., and Hunter, C.L. (2006). Quantitative mass spectrometric multiple reaction monitoring assays for major plasma proteins. Mol. Cell. Proteomics *5*, 573–588.

Anderson, N.L., Anderson, N.G., Haines, L.R., Hardie, D.B., Olafson, R.W., and Pearson, T.W. (2004). Mass spectrometric quantitation of peptides and proteins using Stable Isotope Standards and Capture by Anti-Peptide Antibodies (SISCAPA). J. Proteome Res. *3*, 235–244.

Bauer, J.A., Chakravarthy, A.B., Rosenbluth, J.M., Mi, D., Seeley, E.H., De Matos Granja-Ingram, N., Olivares, M.G., Kelley, M.C., Mayer, I.A., Meszoely, I.M., *et al.* (2010). Identification of markers of taxane sensitivity using proteomic and genomic analyses of breast tumors from patients receiving neoadjuvant paclitaxel and radiation. Clin. Cancer Res. *16*, 681–690.

Brand, J., Haslberger, T., Zolg, W., Pestlin, G., and Palme, S. (2006). Depletion efficiency and recovery of trace markers from a multiparameter immunodepletion column. Proteomics *6*, 3236–3242.

Cheng, J., Xie, H.Y., Xu, X., Wu, J., Wei, X., Su, R., Zhang, W., Lv, Z., Zheng, S., and Zhou, L. (2011). NDRG1 as a biomarker for metastasis, recurrence and of poor prognosis in hepatocellular carcinoma. Cancer Lett. *310*, 35–45.

Dasilva, N., Díez, P., Matarraz, S., Gonzalez-Gonzalez, M., Paradinas, S., Orfao, A., and Fuentes, M. (2012). Biomarker discovery by novel sensors based on nanoproteomics approaches. Sensors (Basel). *12*, 2284–2308.

Di Michele, M., Marcone, S., Cicchillitti, L., Della Corte, A., Ferlini, C., Scambia, G., Donati, M.B., and Rotilio, D. (2010). Glycoproteomics of paclitaxel resistance in human epithelial ovarian cancer cell lines: towards the identification of putative biomarkers. J. Proteomics 73, 879–898.

Gillette, M.A., and Carr, S.A. (2013). Quantitative analysis of peptides and proteins in biomedicine by targeted mass spectrometry. Nat. Methods *10*, 28–34.

Heywood, W., Wang, D., Madgett, T.E., Avent, N.D., Eaton, S., Chitty, L.S., and Mills, K. (2012). The development of a peptide SRM-based tandem mass spectrometry assay for prenatal screening of Down syndrome. J. Proteomics *75*, 3248–3257.

Hocker, J.R., Peyton, M.D., Lerner, M.R., Mitchell, S.L., Lightfoot, S.A., Lander, T.J., Bates-Albers, L.M., Vu, N.T., Hanas, R.J., Kupiec, T.C., *et al.* (2011). Serum discrimination of early-stage lung cancer patients using electrospray-ionization mass spectrometry. Lung Cancer *74*, 206–211.

Holman, S.W., Sims, P.F., and Eyers, C.E. (2012). The use of selected reaction monitoring in quantitative proteomics. Bioanalysis *4*, 1763–1786.

Huttenhain, R., Malmstrom, J., Picotti, P., and Aebersold, R. (2009). Perspectives of targeted mass spectrometry for protein biomarker verification. Curr. Opin. Chem. Biol. *13*, 518–525.

Kakinuma, T., Saeki, H., Tsunemi, Y., Fujita, H., Asano, N., Mitsui, H., Tada, Y., Wakugawa, M., Watanabe, T., Torii, H., *et al.* (2003). Increased serum cutaneous T-cell-attracting chemokine (CCL27) levels in patients with atopic dermatitis and psoriasis vulgaris. J. Allergy Clin. Immunol. *111*, 592–597.

Kaur, P., Rizk, N.M., Ibrahim, S., Younes, N., Uppal, A., Dennis, K., Karve, T., Blakeslee, K., Kwagyan, J., Zirie, M., *et al.* (2012). iTRAQ-based quantitative protein expression profiling and MRM verification of markers in type 2 diabetes. J. Proteome Res. *11*, 5527–5539.

Kikuta, K., Kubota, D., Saito, T., Orita, H., Yoshida, A., Tsuda, H., Suehara, Y., Katai, H., Shimada, Y., Toyama, Y., *et al.* (2012). Clinical proteomics identified ATP-dependent RNA helicase DDX39 as a novel biomarker to predict poor prognosis of patients with gastrointestinal stromal tumor. J. Proteomics *75*, 1089–1098.

Kroksveen, A.C., Aasebo, E., Vethe, H., Van Pesch, V., Franciotta, D., Teunissen, C.E., Ulvik, R.J., Vedeler, C., Myhr, K.M., Barsnes, H., *et al.* (2013). Discovery and initial verification of differentially abundant proteins between multiple sclerosis patients and controls using iTRAQ and SID-SRM. J. Proteomics *78*, 312–325.

Kulasingam, V., and Diamandis, E.P. (2008). Strategies for discovering novel cancer biomarkers through utilization of emerging technologies. Nat. Clin. Pract. *5*, 588–599.

Le Blanc, J.C., Hager, J.W., Ilisiu, A.M., Hunter, C., Zhong, F., and Chu, I. (2003). Unique scanning capabilities of a new hybrid linear ion trap mass spectrometer (Q TRAP) used for high sensitivity proteomics applications. Proteomics *3*, 859–869.

Lee, S.A., Hong, S., Kim, H.J., Lee, S.H., and Yum, H.Y. (2013). Correlation between serum vitamin d level and the severity of atopic dermatitis associated with food sensitization. Allergy. Asthma Immunol. Res. *5*, 207–210.

Li, J., Xie, Z., Shi, L., Zhao, Z., Hou, J., Chen, X., Cui, Z., Xue, P., Cai, T., Wu, P., *et al.* (2012). Purification, identification and profiling of serum amyloid A proteins from sera of advanced-stage cancer patients. J. Chromatogr. Anal. Technol. Biomed. Life Sci. *889–890*, 3–9.

Li, X.H., Li, C., and Xiao, Z.Q. (2011). Proteomics for identifying mechanisms and biomarkers of drug resistance in cancer. J. Proteomics *74*, 2642–2649.

Liu, T., Hossain, M., Schepmoes, A.A., Fillmore, T.L., Sokoll, L.J., Kronewitter, S.R., Izmirlian, G., Shi, T., Qian, W.J., Leach, R.J., *et al.* (2012). Analysis of serum total and free PSA using immunoaffinity depletion coupled to SRM: correlation with clinical immunoassay tests. J. Proteomics *75*, 4747–4757.

Maekawa, M., Misawa, Y., Sotoura, A., Yamaguchi, H., Togawa, M., Ohno, K., Nittono, H., Kakiyama, G., Iida, T., Hofmann, A.F., *et al.* (2013). LC/ESI-MS/MS analysis of urinary 3beta-sulfooxy-7beta-N-acetylglucosaminyl-5-cholen-24-oic acid and its amides: new biomarkers for the detection of Niemann-Pick type C disease. Steroids *78*, 967–972.

Mateos, J., Lourido, L., Fernandez-Puente, P., Calamia, V., Fernandez-Lopez, C., Oreiro, N., Ruiz-Romero, C., and Blanco, F.J. (2012). Differential protein profiling of synovial fluid from rheumatoid arthritis and osteoarthritis patients using LC-MALDI TOF/TOF. J. Proteomics *75*, 2869–2878.

Morofuji, N., Ojima, H., Onaya, H., Okusaka, T., Shimada, K., Sakamoto, Y., Esaki, M., Nara, S., Kosuge, T., Asahina, D., *et al.* (2012). Macrophage-capping protein as a tissue biomarker for prediction of response to gemcitabine treatment and prognosis in cholangiocarcinoma. J. Proteomics *75*, 1577–1589.

Nahnsen, S., and Kohlbacher, O. (2012). In silico design of targeted SRM-based experiments. BMC Bioinformatics *13 Suppl 1*, S8–2105–13–S16–S8. Epub 2012 Nov 5.

Nicol, G.R., Han, M., Kim, J., Birse, C.E., Brand, E., Nguyen, A., Mesri, M., FitzHugh, W., Kaminker, P., Moore, P.A., *et al.* (2008). Use of an immunoaffinity-mass spectrometry-based approach for the quantification of protein biomarkers from serum samples of lung cancer patients. Mol. Cell. Proteomics *7*, 1974–1982.

Nicolini, C., Bragazzi, N., and Pechkova, E. (2012). Nanoproteomics enabling personalized nanomedicine. Adv. Drug Deliv. Rev. *64*, 1522–1531.

Pin, E., Fredolini, C., and Petricoin 3rd, E.F. (2013). The role of proteomics in prostate cancer research: biomarker discovery and validation. Clin. Biochem. *46*, 524–538.

Poschmann, G., Sitek, B., Sipos, B., Hamacher, M., Vonend, O., Meyer, H.E., and Stuhler, K. (2009). Cell-based proteome analysis: the first stage in the pipeline for biomarker discovery. Biochim. Biophys. Acta *1794*, 1309–1316.

Rafalko, A., Iliopoulos, O., Fusaro, V.A., Hancock, W., and Hincapie, M. (2010). Immunoaffinity enrichment and liquid chromatography-selected reaction monitoring mass spectrometry for quantitation of carbonic anhydrase 12 in cultured renal carcinoma cells. Anal. Chem. *82*, 8998–9005.

Ruiz-Romero, C., and Blanco, F.J. (2010). Proteomics role in the search for improved diagnosis, prognosis and treatment of osteoarthritis. Osteoarthritis Cartilage *18*, 500–509.

Sandoval, J.A., Dobrolecki, L.E., Huang, J., Grosfeld, J.L., Hickey, R.J., and Malkas, L.H. (2006). Neuroblastoma detection using serum proteomic profiling: a novel mining technique for cancer? J. Pediatr. Surg. *41*, 639–646.

Schiess, R., Wollscheid, B., and Aebersold, R. (2009). Targeted proteomic strategy for clinical biomarker discovery. Mol. Oncol. 3, 33–44.

Schoenherr, R.M., Zhao, L., Whiteaker, J.R., Feng, L.C., Li, L., Liu, L., Liu, X., and Paulovich, A.G. (2010). Automated screening of monoclonal antibodies for SISCAPA assays using a magnetic bead processor and liquid chromatography-selected reaction monitoring-mass spectrometry. J. Immunol. Methods *353*, 49–61.

Schwamborn, K., Krieg, R.C., Grosse, J., Reulen, N., Weiskirchen, R., Knuechel, R., Jakse, G., and Henkel, C. (2009). Serum proteomic profiling in patients with bladder cancer. Eur. Urol. *56*, 989–996.

Shi, T., Su, D., Liu, T., Tang, K., Camp 2nd, D.G., Qian, W.J., and Smith, R.D. (2012). Advancing the sensitivity of selected reaction monitoring-based targeted quantitative proteomics. Proteomics *12*, 1074–1092.

Somparn, P., Hirankarn, N., Leelahavanichkul, A., Khovidhunkit, W., Thongboonkerd, V., and Avihingsanon, Y. (2012). Urinary proteomics revealed prostaglandin H(2)D-isomerase, not Zn-alpha2-glycoprotein, as a biomarker for active lupus nephritis. J. Proteomics *75*, 3240–3247.

Stahl-Zeng, J., Lange, V., Ossola, R., Eckhardt, K., Krek, W., Aebersold, R., and Domon, B. (2007). High sensitivity detection of plasma proteins by multiple reaction monitoring of N-glycosites. Mol. Cell. Proteomics *6*, 1809–1817.

Tamae, D., Byrns, M., Marck, B., Mostaghel, E.A., Nelson, P.S., Lange, P., Lin, D., Taplin, M.E., Balk, S., Ellis, W., *et al.* (2013). Development, validation and application of a stable isotope dilution liquid chromatography electrospray ionization/selected reaction monitoring/mass spectrometry (SID-LC/ESI/SRM/MS) method for quantification of keto-androgens in human serum. J. Steroid Biochem. Mol. Biol. *138C*, 281–289.

Veenstra, T.D. (2007). Global and targeted quantitative proteomics for biomarker discovery. J. Chromatogr. Anal. Technol. Biomed. Life Sci. *847*, 3–11.

Wei, M., Duan, D., Liu, Y., Wang, Z., and Li, Z. (2013). Increased thymosin beta4 levels in the serum and SF of knee osteoarthritis patients correlate with disease severity. Regul. Pept. *185*, 34–36.

Whiteaker, J.R., Zhao, L., Abbatiello, S.E., Burgess, M., Kuhn, E., Lin, C., Pope, M.E., Razavi, M., Anderson, N.L., Pearson, T.W., *et al.* (2011). Evaluation of large scale quantitative proteomic assay development using peptide affinity-based mass spectrometry. Mol. Cell. Proteomics *10*, M110.005645.

Whiteaker, J.R., Zhao, L., Lin, C., Yan, P., Wang, P., and Paulovich, A.G. (2012). Sequential multiplexed analyte quantification using peptide immunoaffinity enrichment coupled to mass spectrometry. Mol. Cell. Proteomics *11*, M111.015347.

Witebsky, E., Rose, N.R., Terplan, K., Paine, J.R., and Egan, R.W. (1957). Chronic thyroiditis and autoimmunization. J. Am. Med. Assoc. *164*, 1439–1447.

Ye, X., Blonder, J., and Veenstra, T.D. (2009). Targeted proteomics for validation of biomarkers in clinical samples. Brief. Funct. Genomic. Proteomic. *8*, 126–135.

Yocum, A.K., and Chinnaiyan, A.M. (2009). Current affairs in quantitative targeted proteomics: multiple reaction monitoring-mass spectrometry. Brief. Funct. Genomic. Proteomic. *8*, 145–157.

Zhao, Y., and Brasier, A.R. (2013). Applications of selected reaction monitoring (SRM)-mass spectrometry (MS) for quantitative measurement of signaling pathways. Methods *61*, 313–322.

Zhao, L., Whiteaker, J.R., Pope, M.E., Kuhn, E., Jackson, A., Anderson, N.L., Pearson, T.W., Carr, S.A., and Paulovich, A.G. (2011). Quantification of proteins using peptide immunoaffinity enrichment coupled with mass spectrometry. J. Vis. Exp. *(53)*. *pii:*, 10.3791/2812.

Targeted Proteomics: Applications in the Study of Liver Disorders

3

Fernando J. Corrales

Abstract

Liver diseases afflict currently more than 10% of the world population and their incidence is increasing. Although the main risk factors are known and the population at risk is routinely monitored, new biomarkers are urgently needed to allow early diagnosis and more effective therapeutic interventions. As proteins are the cellular tools used to perform most of the biological processes, proteomics, among all the commonly known as -omics technologies, is expected to provide, a closer envisage of cellular functions and seminal information to understand human physiology in health and disease. Large-scale comparative proteomics have provided in the last few years great amounts of data that partially outline the principal pathogenic mechanisms of some liver diseases and point out to potential biomarkers. However, the stringent validation of the emerging hypotheses and the detection of covalently modified protein species of clinical relevance that usually represent a sub-stoichiometric and very low abundant fraction in complex biological matrices are still bottlenecks in translational research. Targeted proteomic approaches based on mass spectrometry and affinity-based enrichment methods will surely contribute to fill the gap between the discovery of large lists of candidate proteins and the evaluation of their clinical use on suitable sample cohorts.

The liver

The liver is the largest organ in the body (1200–1500 g), representing about one-fiftieth of the total adult body weight. Liver architecture is conventionally organized in four lobes, right, left, caudate and quadrate, according to typographic more than functional arguments. From a physiological or functional perspective, the liver is organized in eight segments, numbered from I to VIII, that integrate in larger units or sectors. Each segment has its own independent vascular and biliary systems, facilitating limited segmental resections of the liver without compromising the viability of the organism. Noteworthy, the liver has a unique regenerative capacity to repair itself after a surgical hepatectomy or upon liver injury processes leading to cell death (Fausto, 2000; Taub, 2004). The liver has many complex, essential functions that are often regulated through post-translational modifications of key intermediate proteins. It is a primary regulator of energy metabolism, synthesizes essential proteins, enzymes and cofactors required for digestion and normal bodily function. In addition, it plays a pivotal role in the detoxification and elimination of endogenous and exogenous compounds. The liver receives all blood from the spleen and the digestive tract via the portal vein and receives oxygen-rich blood from the hepatic artery, thereby gaining access to the systemic circulation. Amino acids, carbohydrates, lipids, xenobiotics and other compounds are then taken from the bloodstream and biotransformed to preserve metabolic homeostasis in the body. Hydrophobic waste and toxic substances are also processed and secreted through the production of bile. It is also worth mentioning the relevant immunological function of the liver, as it is the major organ for phagocytosis of circulating material such as foreign or senescent macromolecules and microorganisms. All these biological functions are dependent on the activity and crosstalk between different cell types. There are approximately 202×10^3 cells in each milligram of normal human liver, of which 80% are hepatocytes (about 100 billion in the whole organ), and the other 20% are sinusoidal cells: endothelial, Kupffer, stellate and pit cells.

The hepatocyte is the main functional unit of the liver and is one of the most complex and metabolically diverse cells in the body that performs pivotal functions such as:

1 uptake, storage and regulated release of nutrients into the circulation: carbohydrates, proteins, lipids and vitamins;
2 synthesis of plasma proteins, lipoproteins, glucose, fatty acid, cholesterol and phospholipids;
3 digestion and absorption of dietary fats through the secretion of bile;
4 degradation and detoxification of exogenous and endogenous compounds.

Hepatocytes are different throughout the liver, depending in part on the availability of substrates within incoming blood; it has been shown that the protein complement of hepatocytes varies considerably according to their proximity to the portal (periportal hepatocytes) or to the hepatic veins (pericentral hepatocytes). Oxygen and glucose uptake, and other circulating solutes are preferential in the periportal region (zone 1), where hepatocytes have higher concentrations of enzymes involved in the Krebs cycle, urea synthesis, glutaminase, bile-salt-dependent bile formation and synthesis of some plasma proteins such as albumin or fibrinogen. In contrast, pericentral hepatocytes (adjacent to hepatic veins, zone 3) perform mainly glucose catabolism, production of α_1-antitrypsin or alpha-fetoprotein and xenobiotic biotransformation. Drugs metabolized by P450 enzymes are present in greater amounts in hepatocytes of zone 3, particularly after induction. This metabolic zonation of liver architecture and physiology can contribute to the patterns of damage inflicted on liver by metabolic and physiological stress. Pericentral hepatocytes, for instance, receive higher concentration of any toxic product of drug metabolism and have less glutathione content making them more susceptible to hepatic drug reactions as manifested by the centrilobular necrosis induced by hepatotoxins.

Endothelial cells constitute the wall of liver sinusoids that are sinusoidal blood vessels containing oxygen-rich blood from the hepatic artery and nutrient-rich blood from the portal vein. These cells act as a sieve between the sinusoid and the space of Disse that separate the sinusoid from the hepatocytes and are very active clearing macromolecules and small particles from the circulation by endocytosis (Smedsrød *et al.*, 1990). This endocytic activity is mediated by numerous surface receptors for glycoproteins, connective tissue macromolecules and lipoproteins. Hepatic endothelial cells release a variety of factors, including IL-1, IL-6, interferon and NO, that modulate several hepatocyte activities (Laskin, 1990; Karpen and Crawford, 1999).

Kupffer cells are highly mobile macrophages attached to the endothelial lining of the sinusoid, most abundantly in periportal areas (Kolios *et al.*, 2006). Their main function is removing old and damaged blood cells or cellular debris, bacteria, viruses, parasites and tumour cells by endocytosis including receptor and non-receptor mediated mechanisms. Additionally, they interact cooperatively with hepatocytes in the metabolism of iron and adenosine. Kupffer cells can be activated by a wide range of agents, such as endotoxin, sepsis, shock, interferon-γ, arachidonic acid and tumour necrosis factor (TNF). This activation leads to the production of reactive oxygen species, NO, cytokines, mitogens, prostanoids and others, that might be toxic for the endothelial and surrounding parenchymal cells.

Hepatic stellate cells, lipocytes or Ito cells, lie within the space of Disse and have long cytoplasmic extensions, some in close contact with parenchymal cells. In normal liver they are the major storage site of retinoids. Stellate cells synthesize extracellular matrix proteins and are key mediators of hepatic tissue repair process. Upon hepatocyte injury, these cells transform from quiescent cells to activated myofibroblasts and lose their lipid droplets, proliferate and induce deposition of extracellular matrix during the development of liver cirrhosis.

Pit cells are mobile, liver-specific, natural killer lymphocytes anchored to the sinusoidal surface of the endothelium. They show spontaneous cytotoxicity against tumour and virus-infected hepatocytes.

Shotgun proteomics and liver disorders

Differential analysis of protein expression can be performed by gel-based or gel-free approaches. The most common gel-based method is two-dimensional fluorescence difference gel electrophoresis (2D-DIGE) where proteins from two different samples to be compared are labelled using different fluorescence dyes (Cy2, Cy3, or Cy5), mixed equally and resolved by 2-DE. Scanning at different wavelengths provides multiple images corresponding to different samples and therefore protein levels can be compared within a single gel reducing variability and improving accuracy in protein semiquantitation (Van den Bergh and Arckens, 2004; Tannu and Hemby, 2006). This is a relatively simple method that allows visualization of thousands of proteins, detection of post-translationally modified species, and targeting of protein expression alterations. However, some restrictions are associated to 2-DE preventing a complete description of the proteome. The primary weakness of this method is the limitation for detection of low-abundance proteins, theoretically those at less than 1000 copies per cell (Wilkins *et al.*, 1998; Rabilloud, 2002). Additionally, hydrophobic proteins and those with extremely acidic or basic pI or M_r less than 10 kDa are hardly observed on 2-DE gels. After separation, spots of interest are excised and incubated with proteases, generally trypsin, and the resulted peptides are analysed by mass spectrometric methods to allow protein identification. Tryptic digests can be characterized by matrix-assisted laser desorption/ionization (MALDI) time of flight (TOF) mass spectrometry that provide the m/z ratios of peptide fragments, or peptide mass fingerprint (PMF) of the protein that facilitates its identification by the comparison with the theoretical PMFs of genome-wide protein sequence databanks (Karas and Hillenkamp, 1988). Alternatively, peptides can be separated by liquid chromatography (nanoLC) and analysed by tandem mass spectrometry (MS/MS) that provide information relative to the amino acid sequence (Fenn *et al.*, 1989). The chromatographic column is connected on line with the MS/MS instrument through a nano-electrospray source (ESI) that allows ionization of the eluted material. Once ionized, a particular peptide/ion is isolated and fragmented by collision with an inert gas generating a fragment spectrum that is compared with the theoretically predicted fragments for all peptide sequences in a database to identify the peptide sequence and, accordingly, the precursor protein.

The second major proteomic approach or shotgun proteomics, proposes a gel-free alternative consisting of proteolytic digestion of the protein mixture to produce a large collection of peptides that are then subjected to multidimensional LC and MS/MS analysis. This approach might circumvent some of the typical drawbacks of 2-DE analysis mentioned above and provide means to perform comparative studies by combination of shotgun peptide sequencing and stable isotope labelling (Goshe and Smith, 2003; Tao and Aebersold, 2003). Peptide mixtures from different samples can be chemically [ICAT, isotope-coded affinity tag (Gygi *et al.*, 1999); ITRAQ, isobaric multiplexing tagging system (Thompson *et al.*, 2003; Ross *et al.*, 2004), enzymatically (digestion in $H_2{}^{18}O$) or metabolically [SILAC, stable isotopic labelling with amino acids in cell culture (Conrads *et al.*, 2001)] labelled using light and heavy isotopes, and then combined equally and analysed by nanoLC ESI-MS/MS. In addition to MS-based approaches, immuno-capture screening methods in array formats have been developed to bridge gaps between large and small-scale protein analysis approaches (LaBaer and Ramachandran, 2005; Lueking *et al.*, 2005; Sanchez-Carbayo, 2011; Hause *et al.*, 2011).

Liver disorders, afflicting more than 10% of the world population, are the fifth cause of mortality in the Western countries (Williams, 2006). Human liver is the primary target for damage by hepatitis viruses, many chemicals and drugs. Great efforts have been devoted in recent years to identify driver mechanisms of liver disease progression. Independently of the aetiology, chronic liver injury and inflammation are increasingly recognized as pivotal factors in the advance of the pathogenic process (Bataller and Brenner, 2005). Moreover, the molecular pathogenesis of liver ailments has been extensively analysed using genome-wide approaches (Breuhahn *et al.*, 2011; Rahman *et al.*, 2012; Wang *et al.*, 2013). In this regard, proteomic studies focused on several liver ailments have provided protein profiles that shed light on deregulated cellular pathways and identified candidate biomarkers and therapeutic targets (Santamaria *et al.*, 2008; Uto *et al.*, 2010). However, even though these studies have extended largely our understanding of the mechanisms underlying the onset and progression of liver ailments, we are far from offering an integrated and comprehensive envisage of liver in health and disease. Although recent efforts have been made to configure a systems based interpretation of liver biology (Zhang *et al.*, 2013) to help biologists and clinicians to analyse their data and define novel biomarkers, there are still several constraints precluding the development of clinical applications. Albeit the enormous progress of proteomics technologies that allowed descriptions of liver proteome and its alterations in hepatopathies with unprecedented sensitivity, non-targeted bottom-up strategies leave several unanswered questions mainly related with proteome regulation and dynamics. Moreover, validation of candidate biomarkers and therapeutic targets require analysis of large amount of samples to test for the selected peptides/proteins. Both, the vast extension of the proteome that cannot be studied in its whole complexity, and the necessity of interrogating hundreds of samples for selected candidates, point out the necessity of more selective, targeted strategies. It is also worth noting that translational research needs hypothesis-driven, supervised analysis guided by epidemiological, clinical, or experimental nominators to identify factors modulating disease development or progression. Sample processing approaches to select particular sub-proteomes are an excellent method to reduce the complexity of the sample, increasing the depth of the analysis to get insight into specific questions as some of those discussed in this report that include regulatory aspects by selection and identification of proteins carrying post-translational modifications. Targeted proteomics based on mass spectrometry (selective reaction monitoring, SRM), allowing the absolute quantification of a target protein or group of proteins within a complex sample, has opened new avenues not only for validation processes, but also for the development of analytical methods usable in the clinical practice.

MS-based targeted proteomics: selected reaction monitoring for protein detection and quantitation

High-throughput MS-based proteomics has been embraced by researchers studying liver ailments and applied to a variety of disease conditions (Zhao *et al.*, 2010; Shang *et al.*, 2012; Zhang *et al.*, 2013; Ramakrishnaiah *et al.*, 2013). These studies gave insights into pathogenic principles and open avenues for drug discovery and diagnosis (Diamond *et al.* 2012). However, these wide screening procedures offer a panoramic picture of the relative expression and modification of large numbers of proteins while they are optimal neither for absolute quantitative analysis of predefined targets nor for testing the selected targets across large numbers of samples. SRM has emerged as the method of choice for MS-based targeted proteomics.

SRM allows, theoretically, selective detection and accurate quantitation of any protein in any matrix by measuring precursor to fragment transitions of its proteotypic peptides (PTPs). A PTP is defined as that allowing the unequivocal identification of the target protein. As for other MS-based experiments, biological specimens are processed and protein extracts are trypsin digested. The resulting peptides can be prefractionated (Schäfer *et al.*, 2012) or directly subjected to liquid chromatography (typically reverse phase C_{18} columns) on line with a mass spectrometer, particularly a triple quadrupole that is capable of filtering the selected ions in the first and third quadrupole, with the second serving as collision cell. The first step of a targeted proteomic experiment is the selection of a set of proteins to be monitored in the MS instrument, in a number inversely proportional to the sensitivity and accuracy required in the analysis. Then, specific peptides and transitions must be selected to configure a SRM method that allows specific identification and accurate quantitation (Fig. 3.1). The targeted protein(s) yields tens to hundreds of peptides upon trypsin digestion (Picotti *et al.*, 2007) but only a small subset among them is routinely observed (Kuster *et al.*, 2005). The use of previous information from SRM data and multiple shotgun analyses that are available in databases such as SRM atlas (http://www.srmatlas.org), GPMDB (http://www.thegpm.org) or PRIDE (http://www.ebi.ac.uk/pride/) reduces significantly the time required for the development of the assay. It is essential to ensure that the selected peptides uniquely identify the targeted protein or one isoform thereof. The quantification of a peptide by SRM requires the selection of specific m/z settings for the first and third quadrupoles, which, in combination, results in a highly sensitive and selective detection of the peptide. The m/z value of the first quadrupole is determined by the mass and the predominant charge state of a peptide, while in the third quadrupole, a particular fragment ion of the peptide is selected. As specific m/z values for the parent and fragment ions are specified in the SRM method, modified peptides would not be detected unless specifically targeted (Unwin *et al.*, 2005; Williamson *et al.*, 2006; Mollah *et al.*, 2007; Griffiths *et al.*, 2007).

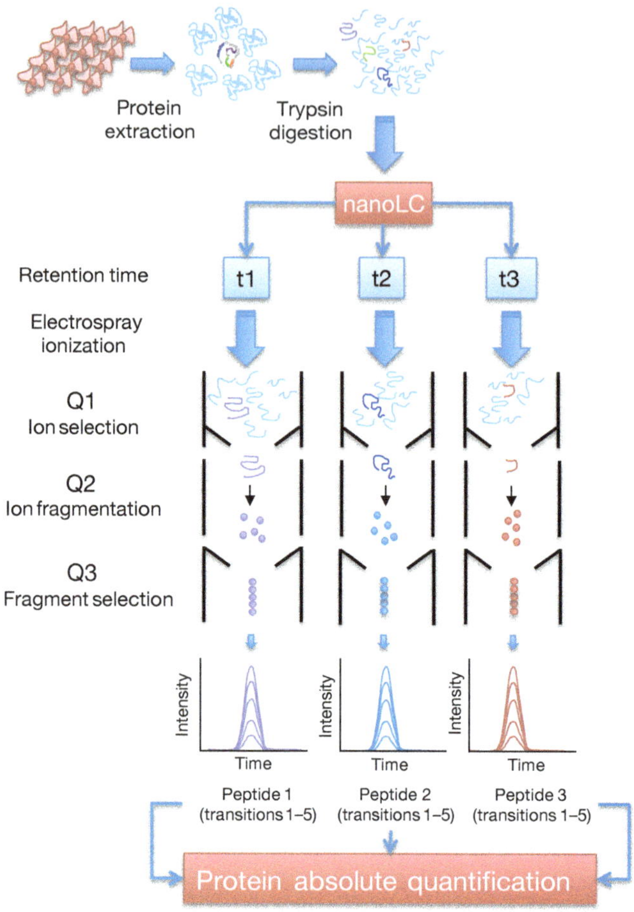

Figure 3.1 Protein quantification by targeted mass spectrometry (SRM). Proteins are extracted from biological matrices and subjected to protease digestion, most commonly trypsin. Tryptic peptides are separated by nano liquid chromatography using reverse phase C_{18} columns connected on line to triple quadrupole mass spectrometers through an electrospray ionization source. Parent ions corresponding to the proteotypic peptides of choice are selected on Q1, fragmented on Q2 and the resulting fragments are selected on Q3. The retention time of the parent to fragment ion *m/z* transitions are monitored. This parameter must be the same for all transitions of a given peptide. The inclusion in the tryptic peptide mixture of stable isotope labelled standard peptides (heavy versions of the proteotypic peptides) allows the absolute quantitation of the protein of interest based on the estimation of 15 independent parameters (3 proteotypic peptides and five transitions in each case) providing an extraordinaire robustness to the method.

Protein quantitation by SRM has a high dynamic range, is absolute rather than relative and has low technical variability when stable isotope labelled standards are spiked in the sample (Gerber *et al.*, 2003; Mallick *et al.*, 2007). These characteristics have led to wide adoption and development of the technique for the targeted quantification of restricted sets of proteins in model systems (Lange *et al.*, 2008; Picotti *et al.*, 2009; Jovanovic *et al.*, 2010), clinical samples (Whiteaker *et al.*, 2007; Schiess *et al.*, 2009; Addona *et al.*, 2009) and toxicology (Jenkins *et al.*, 2006). SRM assays can be easily multiplexed for tens or hundreds of proteins and refined to include novel protein targets or to remove proteins with, for instance, low discriminatory power. Due to all these considerations, SRM assays have become an attractive alternative to antibody-based methods for targeted protein analysis, although SRM assays on the current technical platforms offer lower sensitivities than those based on antibodies.

SRM targeted proteomics is emerging as a reference method for the study of liver disorders, in particular to identify biomarkers in blood that may promote the development of more efficient and non-invasive diagnostic/prognostic tests. One of the complications of chronic liver damage is liver fibrosis. The combination of tissue-specific transcriptomic data with SRM in plasma and serum has been nicely used to identify circulating proteins that allow the stratification of HCV patients in a long-term antiviral trial (Qin *et al.*, 2012). Considering that the concentration of organ-specific proteins in blood can be used to monitor the health status of a specific organ, 130 liver-specific proteins were monitored by targeted SRM assays on a triple quadrupole (QQQ) instrument. Among them, 54 were detected in plasma and 38 in serum; the reduced number likely results from proteolysis derived from the coagulation cascade (Rai *et al.*, 2005). Although stable isotope-labelled standard peptides were used in the study, semiquantitative analysis was performed since depletion of the 14 most abundant proteins was considered as a variation factor relative to protein concentration that preclude absolute quantitation. Interestingly, five proteins were shown to be able to discriminate patients from controls and to efficiently differentiate early fibrosis (Ishak index 2–4) from cirrhosis (Ishak index 5–6). More recently, seven potential biomarkers of liver fibrosis identified on a 2D-electrophoresis study, were subjected to verification by SRM analysis of serum samples from 40 HBV patients, resulting in peptide ions of transferrin as promising candidates for staging liver fibrosis in HBV patients (Xu *et al.*, 2013).

Abusive alcohol consumption is one of the leading causes of chronic liver disease and liver cirrhosis and therefore represents an important social burden (Schwartz and Reinus, 2012; Blachier *et al.*, 2013). The enzyme responsible for the metabolism of ethanol is alcohol dehydrogenase (ADH) and it has been demonstrated that analysis of the content of this enzyme in the liver is important to evaluate the physiological and toxicological effects of alcohol (Bosron *et al.*, 1993; Hurley and Edenberg, 2012). Absolute quantification of ADH in liver appears then an interesting question but it should be bear in mind that there are at least eight isoforms with different ethanol oxidizing capacities that require specific quantitation assays. The use of intact protein isotopically labelled instead of peptides allowed to obtain ADH isoform specific information (Janecki *et al.*, 2007). Moreover, the addition of the intact purified labelled protein to the liver protein extract normalizes the sample preparation procedure and the trypsin digestion step.

The use of internal stable isotope labelled standard peptides is necessary for absolute quantitation in an SRM experiment. These heavy peptides have been usually produced by standard Fmoc chemistry method using isotope-incorporated amino acids. Alternatively, the Qconcat (Rivers *et al.*, 2007) strategy has been proposed to produce labelled peptides through expression of a synthetic gene that encodes a recombinant polypeptide in medium containing the isotopic amino acids. More recently an alternative method has been proposed, consisting of the ^{18}O labelling of standard peptides (Zhao *et al.*, 2010). The specificity and accuracy of the labelling procedure was ensured by optimized tryptic digestion and the SRM method was refined to avoid cross-talk between labelled and non-labelled peptides. The method was successfully applied to measure the levels of two potential biomarkers of HCC, vitronectin and clusterin in serum samples from controls and patients. Interestingly, the concentrations of the two proteins were found to be reduced in hepatocellular carcinoma (HCC) samples to a similar extent as that reported by ELISA (Høgåsen *et al.*, 1993; Boyd *et al.*, 1993).

A serious bottleneck in drug development pipeline are the limitations of current preclinical toxicology to predict early that a drug candidate have to be removed from development or from the market due to toxicity or safety issues (Kola and Landis, 2004). Given the substantial diversity of insults that occur in liver, the study of the hepatic effects induced by model toxicant compounds will help to establish signatures of toxicity. Collins *et al.* (2012) have shown in a recent study that targeted SRM proteomics assays built on predefined biomarker candidates might be an ideal method for large scale verification analysis. As a proof of principle, this study evaluates the effect of the hepatotoxic compound EMD 335823 in rat liver and data from transcriptomic analysis, from label-free LC-MS proteomics discovery and also selected candidates identified from literature were integrated to configure a panel of 48 putative toxicity biomarkers in liver tissue. Interestingly, most of the proteins selected from transcriptomic or label-free experiments were regulated upon treatment, while those selected from literature showed less marked regulation, perhaps due to selections based on different statistical thresholds. Moreover, a reasonable correlation between fold changes determined by label-free, SRM proteomic analysis and transcriptomics, highlighting the value of discovery approaches to develop medium-throughput targeted methods for large-scale verification studies.

Protein glycosylation plays an important role in many central biological functions and, hence, glycoproteins and protein glycosylation have become central aspects of proteomics emerging as a productive source of disease biomarkers. Abnormal protein glycosylation patterns closely associates with cancer progression, including HCC (Miyoshi *et al.*, 2008; Dai *et al.*, 2009). A combination of lectin glycoprotein enrichment and SRM strategy allowed detection of four proteins clearly differentiating HCC from control plasma samples (Ahn *et al.*, 2012). This strategy has shown useful to reduce the difficulty inherent to the identification of low-abundance aberrant protein glycoforms, which are present in sub-stoichiometric amounts with respect the total glycoform of a given protein, and might offer a plausible alternative to immunoassay-based methods for biomarker validation. Particular interest has been paid to haptoglobin as isoforms of this protein carrying site-specific glycosylation changes that correlate with lung (Wang *et al.*, 2011), pancreatic (Miyoshi and Nakano, 2008) and colon (Park *et al.*, 2012) cancers. Site-specific quantitative workflows for the analysis of glycopeptides are not extensively developed due to the complexity of microheterogeneous glycoproteins and to the ionization/fragmentation behaviour of glycopeptides (Zhao *et al.*, 2011). These constrains were circumvented on a recent study where a LC-MS-SRM method was optimized to quantitate haptoglobin-T3 isoforms in patients with liver cirrhosis and HCC (Sanda *et al.*, 2013). In addition to a general increase of fucosylated isoforms in the diseased samples, specific species correlated with MELD score and low platelet count.

Protein regulation: targeted analysis of post-translational modifications based on enrichment procedures

Regulation of protein activity by post-translational modifications (PTM) is a rapid and efficient manner to modulate biological functions, allowing efficient adaptive responses to environmental conditions than ensure cell survival. Many types of PTMs are already known as the result of the covalent bonding of different chemical moieties to specific amino acid side chains (Krishna and Wold, 1993). Proteomics has been the experimental approach of election for massive PTM screening as well as in more targeted analysis. The relative stability of some PTMs and the sub-stoichiometric proportion of the modified species of, in most cases, low-abundance proteins makes its detection and quantification really challenging. Protocols to preserve specific PTMs and enrichment procedures to increase the relative concentrations of the modified proteins are needed to substantially increase the depth of

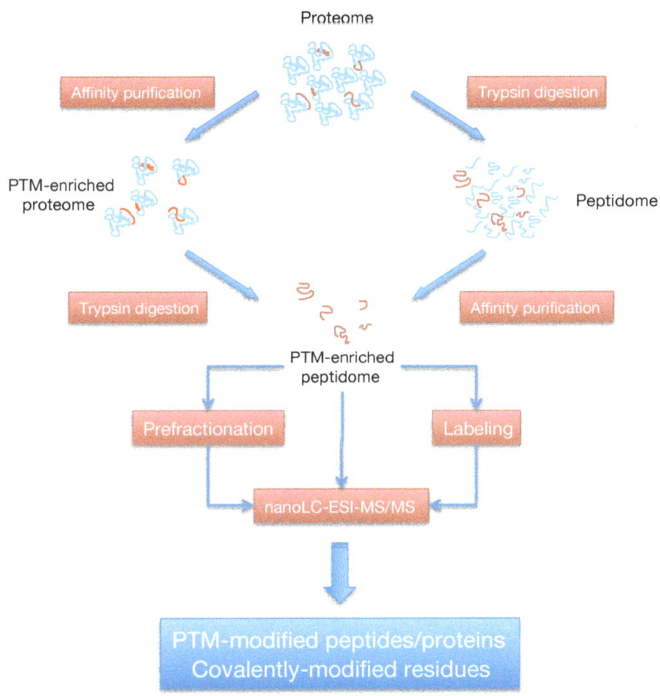

Figure 3.2 General workflow for large-scale analysis of protein post-translational modifications. The identification and characterization of protein PTMs require enrichment procedures to circumvent the low relative abundance of the modified species and thus to enhance the efficiency of the mass spectrometry based analytical methods. The enrichment procedure (based on the use of selective chromatographic moieties or specific antibodies) can be performed directly on the protein mixture resulting on a protein population where all polypeptide ideally carry the PTM under study. Alternatively, the enrichment is done upon tryptic digestion as it is considered that the specificity of the method is increased when peptides instead proteins are selected. Enriched peptides can be then prefractionated, labelled for comparative analyses or directly analysed by reverse phase C18 nano chromatography coupled on-line to a high-resolution mass spectrometer (Orbitap, Qexactive, triple TOF 5600, Sinapt G2, etc.). Modified peptides/proteins are identified according to a shotgun approach and the specific modification sites are determined.

the study (Fig. 3.2). Some of the state of the art methods to target proteomic analysis to specific PTMs have been applied to the study of liver disorders.

Protein phosphorylation

Reversible protein phosphorylation is a complex network of signalling and regulatory events that affects virtually every cellular process. Our understanding of complex phosphorylation networks in the physiological context remains limited due largely to inefficient isolation of serine/threonine phosphorylated peptides. Application of enrichment approaches at the level of phosphorylated proteins and/or peptides using a combination of ion exchange and immobilized metal affinity (IMAC) chromatography have allowed the identification of more than 5000 unique phosphopeptides in liver tissue (Villén *et al.*, 2007; Villén and Gygi, 2008; Demirkan *et al.*, 2012), which configures a detailed map to guide the comprehensive exploration of human liver biological functions regulated by phosphorylation and the changes participating in diseased states or cellular adaptive responses.

Proteasome inhibitors are potential therapeutic agents in the treatment of liver fibrosis and HCC (Ganten *et al.*, 2005; Anan *et al.*, 2006) and this effect is mediated by changes in the phosphorylation state of pivotal regulatory proteins. Phosphoprotein affinity purification and subsequent separation by 2D-DIGE proved its efficiency on identification of proteins differentially phosphorylated such as stathmin and NRDG1 (N-myc down-regulated gene family 1), two proliferation intermediates in liver progenitor-29 cells (Santamaria *et al.*, 2009, 2012). Deeper phosphoproteome analyses have been however achieved using different combinations of phosphorylated protein/peptide enrichment and tandem mass spectrometry. Some anticancer treatments rely on drugs targeting protein kinases that regulate central signalling pathways typically active on tumour cells. One of such drugs is rapamicin, an agent interfering mTOR signalling pathway (Wullschleger *et al.*, 2006). Large-scale phosphoproteomic profiling using a combination of sequential protein and peptide enrichment steps resulted in a detailed description of alterations of mTOR complex1 (a nutrient and growth factor responsive cell signalling network) downstream signalling in rat liver upon rapamicin administration. More than 3000 phosphopeptides were identified, 67 of which were changed by the drug and revealed regulation of proteins associated with translation, mTOR interacting proteins, AMP kinase, among others (Demirkan *et al.*, 2011), providing potential new mTORC1 mediated phosphorylation events.

The liver is a central organ for detoxification a function mediated by the activity of cytochrome P450 (CYP). Understanding the regulatory mechanisms of these proteins is central and is an important issue in the process of drug development. CYP450 superfamily is integrated by haem-dependent enzymes that are membrane-anchored in the endoplasmic reticulum or in the mitochondria. Twelve of the human CYP forms are known to play a major role in the metabolism of xenobiotics, including most clinically used drugs (Lewis, 2004). Eight phosphorylation sites were discovered in P450 by MS analysis in an enriched fraction of a CYP450 from human liver microsomes followed by TiO_2 phosphopeptide enrichment (Redlich *et al.*, 2008). These results suggest regulatory mechanisms of these important enzymes but the functional significance of these reported modifications should be further supported by quantitative analysis upon liver cells stimulation as it has been the case of CYP3A4, responsible for the disposition of over 50% clinically relevant drugs (Wang *et al.*, 2012).

The introduction of stable isotopes allowed quantification of thousands of phospho-sites. Mitochondrial dysfunction is suspected to contribute to the onset of obesity and insulin resistance (Szendroedi *et al.*, 2012). There is increasing evidence supporting that mitochondrial adaptation relies on a range of reversible PTMs, including protein acetylation, oxidation and phosphorylation. This idea finds support on a recent study based on high throughput quantitative (iTRAQ) phosphoproteomic study (SCX and each fraction TiO2, then MS) in which among 4000 unique phospho-sites identified, 692 were mitochondrial, representing 449 phosphoisoforms, and 102 among them displayed a differential regulation in liver from obese mice (lectin deficient). Interestingly, dynamic changes dependent on mouse strain, age, the onset of obesity and the acute fasting/refeeding were demonstrated and revealed adaptation of ketogenesis during the onset of obesity and type 2 diabetes. (Grimsrud *et al.*, 2012). Additional labelling procedures have been developed as alternative to a reductive dimethylation protocol has been used to explore the response of the mouse liver phosphoproteome to fasting-refeeding. Of more than 8000 phospho-sites, above 7000 were reliably quantified, corresponding to over 2300 phosphoproteins, of which roughly 400 were differentially phosphorylated and suggest regulation of central cellular pathways (Wilson-Grady *et al.*, 2013).

Antibody arrays targeting specific proteins have also shown their capacity to correlate phosphorylation events with the progression of non-alcoholic esteatohepatitis (NASH). NASH is the liver manifestation of metabolic syndrome and is linked to insulin resistance, type 2 diabetes and obesity (Sanyal *et al.*, 2001). Its prevalence is increasing in western countries and is a matter of intense investigations as it might progress to liver fibrosis, cirrhosis and HCC (Younossi *et al.*, 2002). Definition of novel parameters helping the stratification of afflicted patients is then of urgent need. The use of customized reverse-phase protein microarrays to assay the phosphorylated and non-phosphorylated isoforms of 27 target proteins in adipose tissue of NASH patients resulted in the identification of differentially regulated insulin receptor and Akt pathways. These data in combination with the clinical parameters allowed an efficient clustering of the studied patients, but the diagnostic/prognostic significance of the method need further investigation (Younossi *et al.*, 2010). It must be understood that alterations in adipose tissue have a direct influence in parenchymal liver cells and this contributes to the setting of NASH. Two soluble biomarkers of NASH, CCL-2 (chemokine C-C motif ligand 2) and Fas ligand, have been also described using an *in silico* selection (knowledge-based approach) of potential candidates that exhibit differential phosphorylation (Page *et al.*, 2013).

Protein acetylation

Protein lysine acetylation is one of the most common reversible post-translational modifications among cellular proteins and regulates a variety of physiological processes including enzyme activity, protein–protein interactions, gene expression, and subcellular localization (Glozak *et al.*, 2005). Therefore, it is a finely tuned process controlled by acetyltransferases and deacetylases, including sirtuins (SIRT) (Schwer and Verdin, 2008), most importantly SIRT3 (Lombard *et al.*, 2007), an acetyl transferase that regulates lysine acetylation of multiple proteins across several metabolic pathways. Like other PTMs, altered acetylation has been related to numerous disease states, including cancer, ageing, cardiovascular, metabolic and alcoholic liver disease (ALD) (Lavu *et al.*, 2008; Alcaín and Villalba, 2009a,b;

Close *et al.*, 2010). This PTM is especially prevalent in mitochondrial proteins, is organ and cell type dependent and involve preferentially enzymes that participate in intermediate metabolism (Choudhary *et al.*, 2009; Zhao *et al.*, 2010). The characterization of the acetyl-proteome has been greatly extended in the last few years by the combination of enrichment of K-acetylated proteins/peptides based on specific antibodies and high sensitivity and accuracy MS instruments.

In addition to histone and nuclear transcription regulators, lysine acetylation is a prevalent modification in enzymes that catalyse intermediate metabolism. Extra nuclear lysine acetylation can be considered as part of a fasted/re-fed feedback control system for the regulation of the metabolic network in fuel switching, where acetyl CoA would be provided for fatty acid oxidation, or glycolysis, respectively. The characterization of protein acetylation patterns is thus of sound interest in conditions wherein normal fuel switching is disrupted, such in type 2 diabetes. According to this, metabolic enzymes are highly represented in mammalian cells among the approximately 2000 reported acetylated proteins (Choudhary *et al.*, 2009; Zhao *et al.*, 2010). It is noteworthy that enzymes involved in glycolysis, gluconeogenesis, the tricalboxylic acid cycle, fatty acid oxidation, the urea cycle, glycogen metabolism, OXPHOS, and amino acid metabolism are acetylated (Guan and Xiong, 2011). These acetylation patterns are organ specific, and importantly affect mitochondrial proteins, as shown in large-scale proteomic studies, even though no mitochondrial enrichment preceded the mass spectrometry analysis (Yang *et al.*, 2011).

SIRT3 appears to be the primary mediator of mitochondrial acetylation (Lombard *et al.*, 2007). In mouse liver, acetylation of 193 proteins, mainly involved in energy metabolism, is increased upon feeding a high-fat diet and correlates with decreased availability of NAD^+ and SIT3 activity (Kendrick *et al.*, 2011). SIRT3 expression is stimulated in the liver upon fasting (Hirschey *et al.*, 2010). Some of its specific targets emerged from large-scale proteomic studies were acetylation patterns of mitochondrial proteins were compared in livers from normal and SIRT3 knockout mice according to an elegant experimental approach that include MS1 filtering for peptide quantitation (Schilling *et al.*, 2012). Among 2187 unique sites corresponding to 483 proteins, lysine acetylation of 283 from 136 proteins were increased in the SIRT3 deficient liver. These results indicate an almost ubiquitous acetylation of proteins in mitochondria, often in multiple sites, suggesting the regulation of multiple metabolic pathways, including ketogenesis, amino acid catabolism, urea and tricarboxylic acid cycles, and particularly fatty acid oxidation, and stand out for the pivotal participation of SIRT3 in the metabolic adaptation of cells to changes in nutrient availability (Rardin *et al.*, 2013). More specifically, SIRT3 regulates ketone body production during fasting by modulating the acetylation status of 3-hydroxy-3-methylglutaryl CoA synthase 2 (HMGCS2). This enzyme catalyses the rate-limiting state of beta-hydroxybutyrate synthesis and its activity is enhanced by de-acetylation of K310, K447 and K473 by SIRT3 (Shimazu *et al.*, 2010).

The liver is the major site of ethanol metabolism, and thus is highly susceptible to alcohol-induced injury. Ethanol consumption induce hyperacetylation of liver mitochondrial proteins with the concomitant imbalance in critical metabolic and antioxidant pathways associated to ALD (Shepard *et al.*, 2010), suggesting an important role of mitochondrial protein acetylation in the pathogenesis of ALD.

Protein glycosylation

Protein glycosylation is the most versatile and common protein modification that plays important roles in various biological processes and disease progression, including cancer (Adamczyk *et al.*, 2012). Experimental evidences support the participation of glycoproteins and glycans in cell–cell and cell–matrix recognition, development and morphogenesis, evasion of immune response as well as other central biological functions (Dell and Morris, 2001; Galonić and Gin, 2007; Gao *et al.*, 2008). Different top-down and bottom-up workflows based on electrophoretic or chromatographic protein separation strategies in combination with MS and glycoprotein enrichment procedures have been proposed to get insights into the human glycoproteome and to discover novel hallmarks of disease (Dai *et al.*, 2009; Pan *et al.*, 2011; Patrie *et al.*, 2013). The identification of glycopeptides in a complex protein digest is rather difficult since glycopeptide signals are often suppressed in the presence of other peptides, especially if the glycan moiety is heterogeneous. Accordingly, enrichment and isolation before MS analysis has become a key issue in glycoproteomics and glycopeptide/glycan-based biomarker discovery. To this end, various strategies have been developed as those based on solid phase methods such as the periodate/hydrazide protocol (Zhang *et al.*, 2003; Tian *et al.*, 2007; Zou *et al.*, 2008), hydrophilic affinity chromatography (Wada *et al.*, 2004), and the capture of glycoproteins with lectins (Kaji *et al.*, 2003; Yang and Hancock, 2004; Pilobello *et al.*, 2007; Fanayan *et al.*, 2012). In contrast to the periodate/hydrazide a method that involve long chemical procedures resulting in the loss of the glycan moiety, lectin affinity chromatography has been extensively used for the isolation of glycoproteins and glycans from various biological sources: plasma (Jung *et al.*, 2009), bile (Kristiansen *et al.*, 2004), urine (Wang *et al.*, 2006) and cells (Xu *et al.*, 2007). These enrichment approaches appear to be complementary as little overlap is observed when applied to the same sample (Cao *et al.*, 2009) and greatly enhance the MS signals of glycopeptides facilitating the sequence and glycosite assignments in typical shotgun analyses (Sun *et al.*, 2007). The subsequent stable isotope labelling of the enriched fraction according to different protocols (Zhang *et al.*, 2003, 2011; Kaji *et al.*, 2000; Sun *et al.*, 2012) allow the robust semi quantitative analysis for a differential glycoproteomic analysis. In addition to the pitfalls of protein glycosylation studies inherent to the enormous complexity, wide dynamic range distribution and low stoichiometry of glycoproteins, large scale analyses of N-glycopeptides by MS is further challenged by their size that is often far from the preferred mass range of shotgun proteomics standards. Digestion with proteases additional to trypsin circumvent partially these limitations resulting from primary structure constraints as well as from the effect of the modifying hydrocarbon moiety that might eventually block the access of trypsin to an adjacent K or R residue (Chen *et al.*, 2009). Taking advantage of the conversion of N to D that occur upon peptide-N-glycanase F (PNGase F) digestion (this amino acid change introduces a mass difference of 0.984 Da), a targeted method to determine N-glycosylation occupancy was developed, in which stable isotope-labelled peptides were synthesized to mimic the deglycosylated form of candidate glycopeptides as internal standards (Pan *et al.*, 2005). Different combinations of these experimental approaches have been delineated to configure optimized workflows aimed at the study of liver glycoproteome to define novel mechanisms of liver disease progression, mainly HCC, and to identify potential biomarkers.

As for other neoplastic disorders, alterations are known to occur during the development of HCC both, at the level of proteins that are glycosylated and also concerning the modifying glycan structures. Since most disease-related proteins secreted from cells to the bloodstream, shed from cell surface or released from tissues, are glycosylated (Gahmberg and Tolvanen, 1996) there is an increasing interest in the study of serum glycoproteome to identify potential biomarkers that might pave the way to the development of more efficient and non-invasive diagnostic/prognostic methods. Lectin-based glycopeptide enrichment followed by treatment with PNGase F of the enriched fraction has been applied to the search of hallmark glycoproteins in HCC (Chaerkady *et al.*, 2008). Moreover, this method allow the isolation of the peptidic and glycan moieties that can be subsequently characterized by MS, as has been beautifully shown by Yang *et al.* (2013) that reported 34 differential glycoproteins and 22 different glycans in the serum of HCC patients taking as reference serum samples from healthy volunteers. A combination of lectin enrichment of serum proteins and isobaric iTRAQ labelling lead to the LC-MS/MS based identification of differential proteins including alpha-1-antitrypsin and Apolipoprotein A1 as potential biomarkers of HCC metastasis (Qin *et al.*, 2013). The availability of lectins with distinct specificities for glycan structures (Dai *et al.*, 2009) facilitates the development of selective affinity chromatography methods that, in addition to the enrichment, provide information relative to the type of structure modifying the protein of interest. Kaji *et al.* (2006) took advantage of this resource and developed a sophisticated and efficient method that allowed the identification of hundreds of glycosylated proteins and their site of glycosylation by a PNGase-mediated incorporation of an ^{18}O stable isotope tag. The potential of this strategy has been outlined in a recent study where roughly 240 branched glycan and 740 fucosylated glycan proteins were found to be differentially represented in the serum of HCC patients, 21 of which were proposed as potential glycobiomarkers of HCC and feasibly of liver fibrosis and cirrhosis (Kaji *et al.*, 2013). Stable isotope ^{18}O carbohydrate labelling using specific endoglycosidases has been proposed as a valuable method to identify differential glycan structures linked to proteins. In total, nine glycans increased in the serum of HCC patients, including biantennary N-glycans and bisecting GlcNAc and NeuNAc residues (Zhang *et al.*, 2011).

Changes in the fucosylation degree of particular serum proteins has been proposed as indicative of liver fibrosis (IgG) and HCC (AFP) progression (Mehta and Block, 2008). Noteworthy, the hyperfucosylated form of AFP, AFP-L3, gained approval from the FDA in 2005 as the only diagnostic assay for HCC. GP73, a resident Golgi membrane protein, was also found to be up-regulated and hyperfucosylated in sera from HCC patients using a strategy that involved the identification of the differential glycan first to then select the carrier proteins by affinity chromatography with the specific lectin (Block *et al.*, 2005). Increased fucosylation of haptoglobin-β relative to the protein levels has been also reported in the serum of patients suffering form HBV-related liver cirrhosis and HCC, suggesting that the ratio of glycosylated to non-glycosylated forms might be a valuable factor for the surveillance of HBV patients (Shu *et al.*, 2011). Moreover, a complex pattern of hypersialylated and hyperfucosylated forms of haptoglobin has been correlated with HCC tumour progression (Ang *et al.*, 2006). Galectin-3 has been also identified as a potential biomarker of HCC based on its differential glycosylation profile. Solid phase based dimethyl labelling revealed increased N-glycosylation of Galectin-3 in five specific N residues in the serum of HCC patients as compared to controls (Sun *et al.*, 2012).

Alternative methods have been also applied in an attempt to define the glycoproteome and its alterations in liver diseases. Development of magnetic nanoparticles engineered with the trapping lectin of election emerges as an efficient enrichment approach in large scale glycoproteomics research of complex biological matrices, as has been demonstrated in HCC serum samples (Yang *et al.*, 2013) and in HCC cell lines (Tang *et al.*, 2010). In contrast to shotgun-based analyses, 2DE Pro-Q emerald staining and protein identification by MS was used to define glycoprotein patterns in three cell lines, one normal (Chang), one non-metastatic (Hep3b) and one highly metastatic (MHCC97H), that revealed differential glycosylation of specific proteins, including AFP (typically used for the diagnosis of HCC), that are proposed as potential biomarkers of hepatocyte transformation and metastatic capacity (Zhou *et al.*, 2007). Cell surface glycoproteins are one of the most frequently observed phenomena paralleling malignant cell growth. The inherent constrains of membrane protein purification and analysis can be circumvented by biotin labelling of surface membrane glycoproteins in intact cells, glycopeptide purification on affinity columns with immobilized avidin and final LC-MS analysis. This strategy leaded to the identification of 119 glycosylation sites from 116 glycopeptides from 79 glycoproteins in human hepatoma HepG2 cell surface, including some proteins that correlate with tumour prognosis and regulate tumour progression and metastasis as CD44, CD147 and EGFR (Mi *et al.*, 2012).

Concluding remarks and future perspectives

The proteomics field has advanced massively in recent years, providing tools to analyse biological systems with unprecedented sensitivity and accuracy. Hundreds to thousands of proteins can be analysed on a single experiment resulting in a precious source of data to understand human biology and disease. Identification of pivotal proteins driving cellular pathways and pathogenic processes shed light into the underlying molecular mechanisms and provided biochemical changes that enable the development of predictive biomarkers that can trigger earlier interventions. Application of proteomic technologies to the study of liver proteome have significantly extended our molecular understanding of liver function but many regulatory and dynamic questions remain still unanswered by these non-targeted approaches. Moreover, validation of candidate biomarkers require selective methods to circumvent the constraints inherent to the vast complexity of the proteome and then tackle the analysis of large cohorts of clinical samples to assess the translational relevance of the differential features identified. Selective procedures to enrich specific subproteomes together with targeted, hypothesis driven analysis are being increasingly used by the biomedical proteomics community and are starting to bridge the gap between analytical and clinical proteomics.

Acknowledgements

The critical reading of the manuscript by Armando Moreno is acknowledged. Financial support was received from ProteoRed and the Carlos III National Health Institute Agreement, ProteoRed-ISCIII; the Agreement between FIMA and the 'UTE project CIMA'; and grants SAF2011-29312 and PIB2010AR-00474 from Ministerio de Ciencia e Innovación.

References

Adamczyk, B., Tharmalingam, T., and Rudd, P.M. (2012). Glycans as cancer biomarkers. Biochim. Biophys. Acta *1820*, 1347–1353.

Addona, T.A., Abbatiello, S.E., Schilling, B., Skates, S.J., Mani, D.R., Bunk, D.M., Spiegelman, C.H., Zimmerman, L.J., Ham, A.-J.L., Keshishian, H., *et al.* (2009). Multi-site assessment of the precision and reproducibility of multiple reaction monitoring-based measurements of proteins in plasma. Nat. Biotechnol. *27*, 633–641.

Ahn, Y.H., Shin, P.M., Oh, N.R., Park, G.W., Kim, H., and Yoo, J.S. (2012). A lectin-coupled, targeted proteomic mass spectrometry (MRM MS) platform for identification of multiple liver cancer biomarkers in human plasma. J. Proteomics *75*, 5507–5515.

Alcaín, F.J., and Villalba, J.M. (2009a). Sirtuin inhibitors. Expert Opin. Ther. Pat. *19*, 283–294.

Alcaín, F.J., and Villalba, J.M. (2009b). Sirtuin activators. Expert Opin. Ther. Pat. *19*, 403–414.

Anan, A., Baskin-Bey, E.S., Bronk, S.F., Werneburg, N.W., Shah, V.H., and Gores, G.J. (2006). Proteasome inhibition induces hepatic stellate cell apoptosis. Hepatology *43*, 335–344.

Ang, I.L., Poon, T.C.W., Lai, P.B.S., Chan, A.T.C., Ngai, S.-M., Hui, A.Y., Johnson, P.J., and Sung, J.J.Y. (2006). Study of serum haptoglobin and its glycoforms in the diagnosis of hepatocellular carcinoma: a glycoproteomic approach. J. Proteome Res. *5*, 2691–2700.

Bataller, R., and Brenner, D.A. (2005). Liver fibrosis. J. Clin. Invest. *115*, 209–218.

Blachier, M., Leleu, H., Peck-Radosavljevic, M., Valla, D.-C., and Roudot-Thoraval, F. (2013). The burden of liver disease in Europe: a review of available epidemiological data. J. Hepatol. *58*, 593–608.

Block, T.M., Comunale, M.A., Lowman, M., Steel, L.F., Romano, P.R., Fimmel, C., Tennant, B.C., London, W.T., Evans, A.A., Blumberg, B.S., *et al.* (2005). Use of targeted glycoproteomics to identify serum glycoproteins that correlate with liver cancer in woodchucks and humans. Proc. Natl. Acad. Sci. U.S.A. *102*, 779–784.

Bosron, W.F., Ehrig, T., and Li, T.K. (1993). Genetic factors in alcohol metabolism and alcoholism. Semin. Liver Dis. *13*, 126–135.

Boyd, N.A., Bradwell, A.R., and Thompson, R.A. (1993). Quantitation of vitronectin in serum: evaluation of its usefulness in routine clinical practice. J. Clin. Pathol. *46*, 1042–1045.

Breuhahn, K., Gores, G., and Schirmacher, P. (2011). Strategies for hepatocellular carcinoma therapy and diagnostics: lessons learned from high throughput and profiling approaches. Hepatology *53*, 2112–2121.

Cao, J., Shen, C., Wang, H., Shen, H., Chen, Y., Nie, A., Yan, G., Lu, H., Liu, Y., and Yang, P. (2009). Identification of N-glycosylation sites on secreted proteins of human hepatocellular carcinoma cells with a complementary proteomics approach. J. Proteome Res. *8*, 662–672.

Chaerkady, R., Thuluvath, P.J., Kim, M.-S., Nalli, A., Vivekanandan, P., Simmers, J., Torbenson, M., and Pandey, A. (2008). O Labeling for a Quantitative Proteomic Analysis of Glycoproteins in Hepatocellular Carcinoma. Clin. Proteomics *4*, 137–155.

Chen, R., Jiang, X., Sun, D., Han, G., Wang, F., Ye, M., Wang, L., and Zou, H. (2009). Glycoproteomics analysis of human liver tissue by combination of multiple enzyme digestion and hydrazide chemistry. J. Proteome Res. *8*, 651–661.

Choudhary, C., Kumar, C., Gnad, F., Nielsen, M.L., Rehman, M., Walther, T.C., Olsen, J.V., and Mann, M. (2009). Lysine acetylation targets protein complexes and co-regulates major cellular functions. Science *325*, 834–840.

Close, P., Creppe, C., Gillard, M., Ladang, A., Chapelle, J.-P., Nguyen, L., and Chariot, A. (2010). The emerging role of lysine acetylation of non-nuclear proteins. Cell Mol. Life Sci. *67*, 1255–1264.

Collins, B.C., Miller, C.A., Sposny, A., Hewitt, P., Wells, M., Gallagher, W.M., and Pennington, S.R. (2012). Development of a pharmaceutical hepatotoxicity biomarker panel using a discovery to targeted proteomics approach. Mol. Cell Proteomics *11*, 394–410.

Conrads, T.P., Alving, K., Veenstra, T.D., Belov, M.E., Anderson, G.A., Anderson, D.J., Lipton, M.S., Pasa-Tolić, L., Udseth, H.R., Chrisler, W.B., *et al.* (2001). Quantitative analysis of bacterial and mammalian proteomes using a combination of cysteine affinity tags and 15N-metabolic labeling. Anal. Chem. *73*, 2132–2139.

Dai, Z., Zhou, J., Qiu, S.-J., Liu, Y.-K., and Fan, J. (2009). Lectin-based glycoproteomics to explore and analyze hepatocellular carcinoma-related glycoprotein markers. Electrophoresis *30*, 2957–2966.

Dell, A., and Morris, H.R. (2001). Glycoprotein structure determination by mass spectrometry. Science *291*, 2351–2356.

Demirkan, G., Yu, K., Boylan, J.M., Salomon, A.R., and Gruppuso, P.A. (2011). Phosphoproteomic profiling of *in vivo* signaling in liver by the mammalian target of rapamycin complex 1 (mTORC1). PLoS ONE *6*, e21729.

Demirkan, G., Salomon, A.R., and Gruppuso, P.A. (2012). Phosphoproteomic analysis of liver homogenates. Methods Mol. Biol. *909*, 151–163.

Diamond, D.L., Krasnoselsky, A.L., Burnum, K.E., Monroe, M.E., Webb-Robertson, B.-J., McDermott, J.E., Yeh, M.M., Dzib, J.F.G., Susnow, N., Strom, S., *et al.* (2012). Proteome and computational analyses reveal new insights into the mechanisms of hepatitis C virus-mediated liver disease posttransplantation. Hepatology *56*, 28–38.

End, F.J., and Friedman, S.L. (2000). Fibrogenesis I. New insights into hepatic stellate cell activation: the simple becomes complex. Am. J. Physiol. Gastrointest. Liver Physiol. *279*, G1–G11.

Fanayan, S., Hincapie, M., and Hancock, W.S. (2012). Using lectins to harvest the plasma/serum glycoproteome. Electrophoresis *33*, 1746–1754.

Fausto, N. (2000). Liver regeneration. J. Hepatol. *32*, 19–31.

Fenn, J.B., Mann, M., Meng, C.K., Wong, S.F., and Whitehouse, C.M. (1989). Electrospray ionization for mass spectrometry of large biomolecules. Science *246*, 64–71.

Gahmberg, C.G., and Tolvanen, M. (1996).Why mammalian cell surface proteins are glycoproteins. Trends Biochem. Sci. *21*, 308–311.

Galonić, D.P., and Gin, D.Y. (2007). Chemical glycosylation in the synthesis of glycoconjugate antitumour vaccines. Nature *446*, 1000–1007.

Ganten, T.M., Koschny, R., Haas, T.L., Sykora, J., Li-Weber, M., Herzer, K., and Walczak, H. (2005). Proteasome inhibition sensitizes hepatocellular carcinoma cells, but not human hepatocytes, to TRAIL. Hepatology *42*, 588–597.

Gao, J., Liu, D., and Wang, Z. (2008). Microarray-based study of carbohydrate-protein binding by gold nanoparticle probes. Anal. Chem. *80*, 8822–8827.

Gerber, S.A., Rush, J., Stemman, O., Kirschner, M.W., and Gygi, S.P. (2003). Absolute quantification of proteins and phosphoproteins from cell lysates by tandem MS. Proc. Natl. Acad. Sci. U.S.A. *100*, 6940–6945.

Glozak, M.A., Sengupta, N., Zhang, X., and Seto, E. (2005). Acetylation and deacetylation of non-histone proteins. Gene *363*, 15–23.

Goshe, M.B., and Smith, R.D. (2003). Stable isotope-coded proteomic mass spectrometry. Curr. Opin. Biotechnol. *14*, 101–109.

Griffiths, J.R., Unwin, R.D., Evans, C.A., Leech, S.H., Corfe, B.M., and Whetton, A.D. (2007). The application of a hypothesis-driven strategy to the sensitive detection and location of acetylated lysine residues. J. Am. Soc. Mass. Spectrom. *18*, 1423–1428.

Grimsrud, P.A., Carson, J.J., Hebert, A.S., Hubler, S.L., Niemi, N.M., Bailey, D.J., Jochem, A., Stapleton, D.S., Keller, M.P., Westphall, M.S., *et al.* (2012). A quantitative map of the liver mitochondrial phosphoproteome reveals post-translational control of ketogenesis. Cell Metab. *16*, 672–683.

Guan, K.-L., and Xiong, Y. (2011). Regulation of intermediary metabolism by protein acetylation. Trends Biochem. Sci. *36*, 108–116.

Gygi, S.P., Rist, B., Gerber, S.A., Turecek, F., Gelb, M.H., and Aebersold, R. (1999). Quantitative analysis of complex protein mixtures using isotope-coded affinity tags. Nat. Biotechnol. *17*, 994–999.

Hause, R.J., Kim, H.-D., Leung, K.K., and Jones, R.B. (2011). Targeted protein-omic methods are bridging the gap between proteomic and hypothesis-driven protein analysis approaches. Expert Rev. Proteomics *8*, 565–575.

Hirschey, M.D., Shimazu, T., Goetzman, E., Jing, E., Schwer, B., Lombard, D.B., Grueter, C.A., Harris, C., Biddinger, S., Ilkayeva, O.R., *et al.* (2010). SIRT3 regulates mitochondrial fatty-acid oxidation by reversible enzyme deacetylation. Nature *464*, 121–125.

Hurley, T.D., and Edenberg, H.J. (2012). Genes encoding enzymes involved in ethanol metabolism. Alcohol Res. *34*, 339–344.

Høgåsen, K., Mollnes, T.E., Tschopp, J., and Harboe, M. (1993). Quantitation of vitronectin and clusterin. Pitfalls and solutions in enzyme immunoassays for adhesive proteins. J. Immunol. Methods *160*, 107–115.

Janecki, D.J., Bemis, K.G., Tegeler, T.J., Sanghani, P.C., Zhai, L., Hurley, T.D., Bosron, W.F., and Wang, M. (2007). A multiple reaction monitoring method for absolute quantification of the human liver alcohol dehydrogenase ADH1C1 isoenzyme. Anal. Biochem. *369*, 18–26.

Jenkins, R.E., Kitteringham, N.R., Hunter, C.L., Webb, S., Hunt, T.J., Elsby, R., Watson, R.B., Williams, D., Pennington, S.R., and Park, B.K. (2006). Relative and absolute quantitative expression profiling of cytochromes P450 using isotope-coded affinity tags. Proteomics *6*, 1934–1947.

Jovanovic, M., Reiter, L., Picotti, P., Lange, V., Bogan, E., Hurschler, B.A., Blenkiron, C., Lehrbach, N.J., Ding, X.C., Weiss, M., *et al.* (2010). A quantitative targeted proteomics approach to validate predicted microRNA targets in C. elegans. Nat. Methods *7*, 837–842.

Jung, K., Cho, W., and Regnier, F.E. (2009). Glycoproteomics of plasma based on narrow selectivity lectin affinity chromatography. J. Proteome Res. *8*, 643–650.

Kaji, H., Saito, H., Yamauchi, Y., Shinkawa, T., Taoka, M., Hirabayashi, J., Kasai, K.-I., Takahashi, N., and Isobe, T. (2003). Lectin affinity capture, isotope-coded tagging and mass spectrometry to identify N-linked glycoproteins. Nat. Biotechnol. *21*, 667–672.

Kaji, H., Yamauchi, Y., Takahashi, N., and Isobe, T. (2006). Mass spectrometric identification of N-linked glycopeptides using lectin-mediated affinity capture and glycosylation site-specific stable isotope tagging. Nat. Protoc. *1*, 3019–3027.

Kaji, H., Ocho, M., Togayachi, A., Kuno, A., Sogabe, M., Ohkura, T., Nozaki, H., Angata, T., Chiba, Y., Ozaki, H., *et al.* (2013). Glycoproteomic discovery of serological biomarker candidates for HCV/HBV infection-associated liver fibrosis and hepatocellular carcinoma. J. Proteome Res. *12*, 2630–2640.

Karas, M., and Hillenkamp, F. (1988). Laser desorption ionization of proteins with molecular masses exceeding 10,000 daltons. Anal. Chem. *60*, 2299–2301.

Karpen, S.J., and Crawford, J.M. (1999). Cellular and molecular biology of the liver. Curr. Opin. Gastroenterol. *15*, 184–191.

Kendrick, A.A., Choudhury, M., Rahman, S.M., McCurdy, C.E., Friederich, M., Van Hove, J.L.K., Watson, P.A., Birdsey, N., Bao, J., Gius, D., *et al.* (2011). Fatty liver is associated with reduced SIRT3 activity and mitochondrial protein hyperacetylation. Biochem. J. *433*, 505–514.

Kola, I., and Landis, J. (2004). Can the pharmaceutical industry reduce attrition rates? Nat. Rev. Drug Discov. *3*, 711–715.

Kolios, G., Valatas, V., and Kouroumalis, E. (2006). Role of Kupffer cells in the pathogenesis of liver disease. World J. Gastroenterol. *12*, 7413–7420.

Krishna, R.G., and Wold, F. (1993). Post-translational modification of proteins. Adv. Enzymol. Relat. Areas Mol. Biol. *67*, 265–298.

Kristiansen, T.Z., Bunkenborg, J., Gronborg, M., Molina, H., Thuluvath, P.J., Argani, P., Goggins, M.G., Maitra, A., and Pandey, A. (2004). A proteomic analysis of human bile. Mol. Cell Proteomics *3*, 715–728.

Kuster, B., Schirle, M., Mallick, P., and Aebersold, R. (2005). Scoring proteomes with proteotypic peptide probes. Nat. Rev. Mol. Cell Biol. *6*, 577–583

LaBaer, J., and Ramachandran, N. (2005). Protein microarrays as tools for functional proteomics. Curr. Opin. Chem. Biol. *9*, 14–19.

Lange, V., Malmström, J.A., Didion, J., King, N.L., Johansson, B.P., Schäfer, J., Rameseder, J., Wong, C.-H., Deutsch, E.W., Brusniak, M.-Y., *et al.* (2008). Targeted quantitative analysis of Streptococcus pyogenes virulence factors by multiple reaction monitoring. Mol. Cell Proteomics *7*, 1489–1500.

Laskin, D.L. (1990). Nonparenchymal cells and hepatotoxicity. Semin. Liver Dis. *10*, 293–304.

Lavu, S., Boss, O., Elliott, P.J., and Lambert, P.D. (2008). Sirtuins – novel therapeutic targets to treat age-associated diseases. Nat. Rev. Drug Discov. *7*, 841–853.

Lewis, D.F.V. (2004). 57 varieties: the human cytochromes P450. Pharmacogenomics *5*, 305–318.

Lombard, D.B., Alt, F.W., Cheng, H.-L., Bunkenborg, J., Streeper, R.S., Mostoslavsky, R., Kim, J., Yancopoulos, G., Valenzuela, D., Murphy, A., *et al.* (2007). Mammalian Sir2 homolog SIRT3 regulates global mitochondrial lysine acetylation. Mol. Cell Biol. *27*, 8807–8814.

Lueking, A., Cahill, D.J., and Müllner, S. (2005). Protein biochips: A new and versatile platform technology for molecular medicine. Drug Discov. Today *10*, 789–794.

Mallick, P., Schirle, M., Chen, S.S., Flory, M.R., Lee, H., Martin, D., Ranish, J., Raught, B., Schmitt, R., Werner, T., *et al.* (2007). Computational prediction of proteotypic peptides for quantitative proteomics. Nat. Biotechnol. *25*, 125–131.

Mehta, A., and Block, T.M. (2008). Fucosylated glycoproteins as markers of liver disease. Dis. Markers *25*, 259–265.

Mi, W., Jia, W., Zheng, Z., Wang, J., Cai, Y., Ying, W., and Qian, X. (2012). Surface glycoproteomic analysis of hepatocellular carcinoma cells by affinity enrichment and mass spectrometric identification. Glycoconj. J. *29*, 411–424.

Miyoshi, E., and Nakano, M. (2008). Fucosylated haptoglobin is a novel marker for pancreatic cancer: detailed analyses of oligosaccharide structures. Proteomics *8*, 3257–3262.

Miyoshi, E., Moriwaki, K., and Nakagawa, T. (2008). Biological function of fucosylation in cancer biology. J. Biochem. *143*, 725–729.

Mollah, S., Wertz, I.E., Phung, Q., Arnott, D., Dixit, V.M., and Lill, J.R. (2007). Targeted mass spectrometric strategy for global mapping of ubiquitination on proteins. Rapid Commun. Mass. Spectrom. *21*, 3357–3364.

Page, S., Birerdinc, A., Estep, M., Stepanova, M., Afendy, A., Petricoin, E., Younossi, Z., Chandhoke, V., and Baranova, A. (2013). Knowledge-based identification of soluble biomarkers: hepatic fibrosis in NAFLD as an example. PLoS ONE *8*, e56009.

Pan, S., Zhang, H., Rush, J., Eng, J., Zhang, N., Patterson, D., Comb, M.J., and Aebersold, R. (2005). High throughput proteome screening for biomarker detection. Mol. Cell Proteomics *4*, 182–190.

Pan, S., Chen, R., Aebersold, R., and Brentnall, T.A. (2011). Mass spectrometry based glycoproteomics – from a proteomics perspective. Mol. Cell Proteomics *10*, R110.003251.

Park, S.-Y., Lee, S.-H., Kawasaki, N., Itoh, S., Kang, K., Hee Ryu, S., Hashii, N., Kim, J.-M., Kim, J.-Y., and Hoe Kim, J. (2012). α1-3/4 fucosylation at Asn 241 of β-haptoglobin is a novel marker for colon cancer: a combinatorial approach for development of glycan biomarkers. Int. J. Cancer *130*, 2366–2376.

Patrie, S.M., Roth, M.J., and Kohler, J.J. (2013). Introduction to glycosylation and mass spectrometry. Methods Mol. Biol. *951*, 1–17.

Picotti, P., Aebersold, R., and Domon, B. (2007). The implications of proteolytic background for shotgun proteomics. Mol. Cell Proteomics *6*, 1589–1598.

Picotti, P., Bodenmiller, B., Mueller, L.N., Domon, B., and Aebersold, R. (2009). Full dynamic range proteome analysis of S. cerevisiae by targeted proteomics. Cell *138*, 795–806.

Pilobello, K.T., Slawek, D.E., and Mahal, L.K. (2007). A ratiometric lectin microarray approach to analysis of the dynamic mammalian glycome. Proc. Natl. Acad. Sci. U.S.A. *104*, 11534–11539.

Qin, S., Zhou, Y., Lok, A.S., Tsodikov, A., Yan, X., Gray, L., Yuan, M., Moritz, R.L., Galas, D., Omenn, G.S., and Hood, L. (2012). SRM targeted proteomics in search for biomarkers of HCV-induced progression of fibrosis to cirrhosis in HALT-C patients. Proteomics *12*, 1244–1252.

Qin, X., Chen, Q., Sun, C., Wang, C., Peng, Q., Xie, L., Liu, Y., and Li, S. (2013). High-throughput screening of tumor metastatic-related differential glycoprotein in hepatocellular carcinoma by iTRAQ combines lectin-related techniques. Med. Oncol. *30*, 420.

Rabilloud, T. (2002). Two-dimensional gel electrophoresis in proteomics: old, old fashioned, but it still climbs up the mountains. Proteomics *2*, 3–10.

Rahman, W., Huang, P., Belov, L., Chrisp, J.S., Christopherson, R.I., Stapelberg, P.M., Warner, F.J., George, J., Bowen, D.G., Strasser, S.I., *et al.* (2012). Analysis of human liver disease using a cluster of differentiation (CD) antibody microarray. Liver Int. *32*, 1527–1534.

Rai, A.J., Gelfand, C.A., Haywood, B.C., Warunek, D.J., Yi, J., Schuchard, M.D., Mehigh, R.J., Cockrill, S.L., Scott, G.B.I., Tammen, H., *et al.* (2005). HUPO Plasma Proteome Project specimen collection and handling: towards the standardization of parameters for plasma proteome samples. Proteomics *5*, 3262–3277.

Ramakrishnaiah, V., Thumann, C., Fofana, I., Habersetzer, F., Pan, Q., de Ruiter, P.E., Willemsen, R., Demmers, J.A.A., Stalin Raj, V., Jenster, G., *et al.* (2013). Exosome-mediated transmission of hepatitis C virus between human hepatoma Huh7.5 cells. Proc. Natl. Acad. Sci. U.S.A. *110*, 13109–13113.

Rardin, M.J., Newman, J.C., Held, J.M., Cusack, M.P., Sorensen, D.J., Li, B., Schilling, B., Mooney, S.D., Kahn, C.R., Verdin, E., and Gibson, B.W. (2013). Label-free quantitative proteomics of the lysine acetylome in mitochondria identifies substrates of SIRT3 in metabolic pathways. Proc. Natl. Acad. Sci. U.S.A. *110*, 6601–6606.

Redlich, G., Zanger, U.M., Riedmaier, S., Bache, N., Giessing, A.B.M., Eisenacher, M., Stephan, C., Meyer, H.E., Jensen, O.N., and Marcus, K. (2008). Distinction between human cytochrome P450 (CYP) isoforms and identification of new phosphorylation sites by mass spectrometry. J. Proteome Res. *7*, 4678–4688.

Rivers, J., Simpson, D.M., Robertson, D.H.L., Gaskell, S.J., and Beynon, R.J. (2007). Absolute multiplexed quantitative analysis of protein expression during muscle development using QconCAT. Mol. Cell Proteomics *6*, 1416–1427.

Ross, P.L., Huang, Y.N., Marchese, J.N., Williamson, B., Parker, K., Hattan, S., Khainovski, N., Pillai, S., Dey, S., Daniels, S., *et al.* (2004). Multiplexed protein quantitation in Saccharomyces cerevisiae using amine-reactive isobaric tagging reagents. Mol. Cell Proteomics *3*, 1154–1169.

Sanchez-Carbayo, M. (2011). Antibody microarrays as tools for biomarker discovery. Methods Mol. Biol. *785*, 159–182.

Sanda, M., Pompach, P., Brnakova, Z., Wu, J., Makambi, K., and Goldman, R. (2013). Quantitative liquid chromatography-mass spectrometry-multiple reaction monitoring (LC-MS-MRM) analysis of site-specific glycoforms of haptoglobin in liver disease. Mol. Cell Proteomics *12*, 1294–1305.

Santamaria, E., Muñoz, J., Fernández-Irigoyen, J., Sesma, L., Odriozola, L., and Corrales, F.J. (2008). Proteomic analysis of liver diseases: molecular mechanisms and biomarker discovery. Mini. Rev. Med. Chem. *8*, 1–9.

Santamaria, E., Mora, M.I., Muñoz, J., Sánchez-Quiles, V., Fernández-Irigoyen, J., Prieto, J., and Corrales, F.J. (2009). Regulation of stathmin phosphorylation in mouse liver progenitor-29 cells during proteasome inhibition. Proteomics *9*, 4495–4506.

Santamaria, E., Sánchez-Quiles, V., Fernández-Irigoyen, J., and Corrales, F.J. (2012). A combination of affinity chromatography, 2D DIGE, and mass spectrometry to analyze the phosphoproteome of liver progenitor cells. Methods Mol. Biol. *909*, 165–180.

Sanyal, A.J., Campbell-Sargent, C., Mirshahi, F., Rizzo, W.B., Contos, M.J., Sterling, R.K., Luketic, V.A., Shiffman, M.L., and Clore, J.N. (2001). Nonalcoholic steatohepatitis: association of insulin resistance and mitochondrial abnormalities. Gastroenterology *120*, 1183–1192.

Schäfer, A., Toerne von, C., Becker, S., Sarioglu, H., Neschen, S., Kahle, M., Hauck, S.M., and Ueffing, M. (2012). Two-dimensional peptide separation improving sensitivity of selected reaction monitoring-based quantitative proteomics in mouse liver tissue: comparing off-gel electrophoresis and strong cation exchange chromatography. Anal. Chem. *84*, 8853–8862.

Schiess, R., Wollscheid, B., and Aebersold, R. (2009). Targeted proteomic strategy for clinical biomarker discovery. Mol. Oncol. *3*, 33–44.

Schilling, B., Rardin, M.J., MacLean, B.X., Zawadzka, A.M., Frewen, B.E., Cusack, M.P., Sorensen, D.J., Bereman, M.S., Jing, E., Wu, C.C., *et al.* (2012). Platform-independent and label-free quantitation of proteomic data using MS1 extracted ion chromatograms in skyline: application to protein acetylation and phosphorylation. Mol. Cell Proteomics *11*, 202–214.

Schwartz, J.M., and Reinus, J.F. (2012). Prevalence and natural history of alcoholic liver disease. Clin. Liver Dis. *16*, 659–666.

Schwer, B., and Verdin, E. (2008). Conserved metabolic regulatory functions of sirtuins. Cell Metab. *7*, 104–112.

Shang, S., Plymoth, A., Ge, S., Feng, Z., Rosen, H.R., Sangrajrang, S., Hainaut, P., Marrero, J.A., and Beretta, L. (2012). Identification of osteopontin as a novel marker for early hepatocellular carcinoma. Hepatology *55*, 483–490.

Shepard, B.D., Tuma, D.J., and Tuma, P.L. (2010). Chronic ethanol consumption induces global hepatic protein hyperacetylation. Alcohol Clin. Exp. Res. *34*, 280–291.

Shimazu, T., Hirschey, M.D., Hua, L., Dittenhafer-Reed, K.E., Schwer, B., Lombard, D.B., Li, Y., Bunkenborg, J., Alt, F.W., Denu, J.M., *et al.* (2010). SIRT3 deacetylates mitochondrial 3-hydroxy-3-methylglutaryl CoA synthase 2 and regulates ketone body production. Cell Metab. *12*, 654–661.

Shu, H., Zhang, S., Kang, X., Li, S., Qin, X., Sun, C., Lu, H., and Liu, Y. (2011). Protein expression and fucosylated glycans of the serum haptoglobin-{beta} subunit in hepatitis B virus-based liver diseases. Acta Biochim. Biophys. Sin. (Shanghai) *43*, 528–534.

Smedsrød, B., Pertoft, H., Gustafson, S., and Laurent, T.C. (1990). Scavenger functions of the liver endothelial cell. Biochem. J. *266*, 313–327.

Sun, B., Ranish, J.A., Utleg, A.G., White, J.T., Yan, X., Lin, B., and Hood, L. (2007). Shotgun glycopeptide capture approach coupled with mass spectrometry for comprehensive glycoproteomics. Mol. Cell Proteomics *6*, 141–149.

Sun, Z., Qin, H., Wang, F., Cheng, K., Dong, M., Ye, M., and Zou, H. (2012). Capture and dimethyl labeling of glycopeptides on hydrazide beads for quantitative glycoproteomics analysis. Anal. Chem. *84*, 8452–8456.

Szendroedi, J., Phielix, E., and Roden, M. (2012). The role of mitochondria in insulin resistance and type 2 diabetes mellitus. Nat. Rev. Endocrinol. *8*, 92–103.

Tang, J., Liu, Y., Yin, P., Yao, G., Yan, G., Deng, C., and Zhang, X. (2010). Concanavalin A-immobilized magnetic nanoparticles for selective enrichment of glycoproteins and application to glycoproteomics in hepatocelluar carcinoma cell line. Proteomics *10*, 2000–2014.

Tannu, N.S., and Hemby, S.E. (2006). Two-dimensional fluorescence difference gel electrophoresis for comparative proteomics profiling. Nat. Protoc. *1*, 1732–1742.

Tao, W.A., and Aebersold, R. (2003). Advances in quantitative proteomics via stable isotope tagging and mass spectrometry. Curr. Opin. Biotechnol. *14*, 110–118

Taub, R. (2004). Liver regeneration: from myth to mechanism. Nat. Rev. Mol. Cell Biol. *5*, 836–847.

Thompson, A., Schäfer, J., Kuhn, K., Kienle, S., Schwarz, J., Schmidt, G., Neumann, T., Johnstone, R., Mohammed, A.K.A., and Hamon, C. (2003). Tandem mass tags: a novel quantification strategy for comparative analysis of complex protein mixtures by MS/MS. Anal. Chem. *75*, 1895–1904.

Tian, Y., Zhou, Y., Elliott, S., Aebersold, R., and Zhang, H. (2007). Solid-phase extraction of N-linked glycopeptides. Nat. Protoc. *2*, 334–339.

Unwin, R.D., Griffiths, J.R., Leverentz, M.K., Grallert, A., Hagan, I.M., and Whetton, A.D. (2005). Multiple reaction monitoring to identify sites of protein phosphorylation with high sensitivity. Mol. Cell Proteomics *4*, 1134–1144.

Uto, H., Kanmura, S., Takami, Y., and Tsubouchi, H.(2010). Clinical proteomics for liver disease: a promising approach for discovery of novel biomarkers. Proteome Sci. *8*, 70.

Van den Bergh, G., and Arckens, L. (2004). Fluorescent two-dimensional difference gel electrophoresis unveils the potential of gel-based proteomics. Curr. Opin. Biotechnol. *15*, 38–43.

Villén, J., and Gygi, S.P. (2008). The SCX/IMAC enrichment approach for global phosphorylation analysis by mass spectrometry. Nat. Protoc. *3*, 1630–1638.

Villén, J., Beausoleil, S.A., Gerber, S.A., and Gygi, S.P. (2007). Large-scale phosphorylation analysis of mouse liver. Proc. Natl. Acad. Sci. U.S.A. *104*, 1488–1493.

Wada, Y., Tajiri, M., and Yoshida, S. (2004). Hydrophilic affinity isolation and MALDI multiple-stage tandem mass spectrometry of glycopeptides for glycoproteomics. Anal. Chem. *76*, 6560–6565.

Wang, D., Hincapie, M., Rejtar, T., and Karger, B.L. (2011). Ultrasensitive characterization of site-specific glycosylation of affinity-purified haptoglobin from lung cancer patient plasma using 10 μm i.d. porous layer open tubular liquid chromatography-linear ion trap collision-induced dissociation/electron transfer dissociation mass spectrometry. Anal. Chem. *83*, 2029–2037.

Wang, L., Li, F., Sun, W., Wu, S., Wang, X., Zhang, L., Zheng, D., Wang, J., and Gao, Y. (2006). Concanavalin A-captured glycoproteins in healthy human urine. Mol. Cell Proteomics *5*, 560–562.

Wang, X., Zhang, A., and Sun, H. (2013). Power of metabolomics in diagnosis and biomarker discovery of hepatocellular carcinoma. Hepatology *57*, 2072–2077.

Wang, Y., Guan, S., Acharya, P., Liu, Y., Thirumaran, R.K., Brandman, R., Schuetz, E.G., Burlingame, A.L., and Correia, M.A. (2012). Multisite phosphorylation of human liver cytochrome P450 3A4 enhances Its gp78- and CHIP-mediated ubiquitination: a pivotal role of its Ser-478 residue in the gp78-catalyzed reaction. Mol. Cell Proteomics *11*, M111.010132.

Whiteaker, J.R., Zhang, H., Zhao, L., Wang, P., Kelly-Spratt, K.S., Ivey, R.G., Piening, B.D., Feng, L.-C., Kasarda, E., Gurley, K.E., *et al.* (2007). Integrated pipeline for mass spectrometry-based discovery and confirmation of biomarkers demonstrated in a mouse model of breast cancer. J. Proteome Res. *6*, 3962–3975.

Wilkins, M.R., Gasteiger, E., Sanchez, J.C., Bairoch, A., and Hochstrasser, D.F. (1998). Two-dimensional gel electrophoresis for proteome projects: the effects of protein hydrophobicity and copy number. Electrophoresis *19*, 1501–1505.

Williams, R. (2006). Global challenges in liver disease. Hepatology *44*, 521–526.

Williamson, B.L., Marchese, J., and Morrice, N.A. (2006). Automated identification and quantification of protein phosphorylation sites by LC/MS on a hybrid triple quadrupole linear ion trap mass spectrometer. Mol. Cell Proteomics *5*, 337–346.

Wilson-Grady, J.T., Haas, W., and Gygi, S.P. (2013). Quantitative comparison of the fasted and re-fed mouse liver phosphoproteomes using lower pH reductive dimethylation. Methods *61*, 277–286.

Wullschleger, S., Loewith, R., and Hall, M.N. (2006). TOR signaling in growth and metabolism. Cell *124*, 471–484.

Xu, M.-Y., Qu, Y., Jia, X.-F., Wang, M.-L., Liu, H., Wang, X.-P., Zhang, L.-J., and Lu, L.-G. (2013). Serum proteomic MRM identify peptide ions of transferrin as new fibrosis markers in chronic hepatitis B. Biomed. Pharmacother. *67*, 561–567.

Xu, Z., Zhou, X., Lu, H., Wu, N., Zhao, H., Zhang, L., Zhang, W., Liang, Y.L., Wang, L., Liu, Y., *et al.* (2007). Comparative glycoproteomics based on lectins affinity capture of N-linked glycoproteins from human Chang liver cells and MHCC97-H cells. Proteomics *7*, 2358–2370.

Yang, G., Chu, W., Zhang, H., Sun, X., Cai, T., Dang, L., Wang, Q., Yu, H., Zhong, Y., Chen, Z., *et al.* (2013). Isolation and identification of mannose-binding proteins and estimation of their abundance in sera from hepatocellular carcinoma patients. Proteomics *13*, 878–892.

Yang, L., Vaitheesvaran, B., Hartil, K., Robinson, A.J., Hoopmann, M.R., Eng, J.K., Kurland, I.J., and Bruce, J.E. (2011). The fasted/fed mouse metabolic acetylome: N6-acetylation differences suggest acetylation coordinates organ-specific fuel switching. J. Proteome Res. *10*, 4134–4149.

Yang, Z., and Hancock, W.S. (2004). Approach to the comprehensive analysis of glycoproteins isolated from human serum using a multi-lectin affinity column. J. Chromatogr. A *1053*, 79–88.

Yao, X., Freas, A., Ramirez, J., Demirev, P.A., and Fenselau, C. (2001). Proteolytic ^{18}O labeling for comparative proteomics: model studies with two serotypes of adenovirus. Anal. Chem. *73*, 2836–2842.

Younossi, Z.M., Diehl, A.M., and Ong, J.P. (2002). Nonalcoholic fatty liver disease: an agenda for clinical research. Hepatology *35*, 746–752.

Younossi, Z.M., Baranova, A., Stepanova, M., Page, S., Calvert, V.S., Afendy, A., Goodman, Z., Chandhoke, V., Liotta, L., and Petricoin, E. (2010). Phosphoproteomic biomarkers predicting histologic nonalcoholic steatohepatitis and fibrosis. J. Proteome Res. *9*, 3218–3224.

Zhang, B., Wang, Z., Deng, B., Wu, X., Liu, J., and Feng, X. (2013). Identification of Enolase 1 and Thrombospondin-1 as serum biomarkers in HBV hepatic fibrosis by proteomics. Proteome Sci. *11*, 30.

Zhang, H., Li, X.-J., Martin, D.B., and Aebersold, R. (2003). Identification and quantification of N-linked glycoproteins using hydrazide chemistry, stable isotope labeling and mass spectrometry. Nat. Biotechnol. *21*, 660–666.

Zhang, W., Wang, H., Tang, H., and Yang, P. (2011). Endoglycosidase-mediated incorporation of 18O into glycans for relative glycan quantitation. Anal. Chem. *83*, 4975–4981.

Zhang, Y., Yang, C., Wang, S., Chen, T., Li, M., Wang, X., Li, D., Wang, K., Ma, J., Wu, S., *et al.* (2013). LiverAtlas: a unique integrated knowledge database for systems-level research of liver and hepatic disease. Liver Int. *33*, 1239–1248.

Zhao, S., Xu, W., Jiang, W., Yu, W., Lin, Y., Zhang, T., Yao, J., Zhou, L., Zeng, Y., Li, H., *et al.* (2010). Regulation of cellular metabolism by protein lysine acetylation. Science *327*, 1000–1004.

Zhao, Y., Jia, W., Sun, W., Jin, W., Guo, L., Wei, J., Ying, W., Zhang, Y., Xie, Y., Jiang, Y., *et al.* (2010). Combination of improved (18)O incorporation and multiple reaction monitoring: a universal strategy for absolute quantitative verification of serum candidate biomarkers of liver cancer. J. Proteome Res. *9*, 3319–3327.

Zhao, Y., Jia, W., Wang, J., Ying, W., Zhang, Y., and Qian, X. (2011). Fragmentation and site-specific quantification of core fucosylated glycoprotein by multiple reaction monitoring-mass spectrometry. Anal. Chem. *83*, 8802–8809.

Zhou, H., Liu, Y., Chui, J., Guo, K., Shun, Q., Lu, W., Jin, H., Wei, L., and Yang, P. (2007). Investigation on glycosylation patterns of proteins from human liver cancer cell lines based on the multiplexed proteomics technology. Arch. Biochem. Biophys. *459*, 70–78.

Zou, Z., Ibisate, M., Zhou, Y., Aebersold, R., Xia, Y., and Zhang, H. (2008). Synthesis and evaluation of superparamagnetic silica particles for extraction of glycopeptides in the microtiter plate format. Anal. Chem. *80*, 1228–1234.

Targeted Proteomics for Chronic Lymphocytic Leukaemia

4

Rafael Góngora, Paula Díez, Nieves Ibarrola,
Rosa M. Dégano, Alberto Orfao and Manuel Fuentes

Abstract

Chronic lymphocytic leukaemia (CLL) is the most common leukaemia in Western countries. It is a blood and bone marrow disease that progresses slowly. Traditionally, flow cytometry has been the diagnosis tool of choice due to the wide knowledge in markers field. Other strategies, such as fluorescent *in situ* hybridization, have allowed chromosomal studies related to CLL. Numerous markers such as Zap70 protein or CD38 have been recently established as possible identifiers of the disease. For this, the need of proteomics approaches is evident. In this chapter we introduce basic information about the disease and its features, as well as the most relevant strategies in proteomics area leading to a better understanding of CLL.

Introduction

The overwhelming size and complexity of human proteome requires of methodologies to carry out fast analysis. Although the significant advances in molecular biology and the original genetic tools, only a small part of the proteome has been studied at the biochemical level. After genomic era, proteomics seems to be the real key to face human pathologies and generate a protein expression pattern from each single patient, which allows us to be situated in personalized medicine (Dasilva *et al.*, 2012; Díez *et al.*, 2012; Gonzalez-Gonzalez *et al.*, 2012; Matarraz *et al.*, 2011).

Over the last decade, proteomics has suffered a deep impulse driving the scientific investigation from the genomics to the proteomics era. Hence, proteomics has undergone remarkable progress thanks to the technical advances made in the area; such as the development of high-throughput methodologies which improve and accelerate sequencing techniques and mass spectrometry and enable the study of thousands of proteins in a single experiment with a little amount of sample. Overall, all these advantages have led to the implementation of proteomics as a useful and accurate technique in clinical routine (Dasilva *et al.*, 2012; Díez *et al.*, 2012; Gonzalez-Gonzalez *et al.*, 2012; Matarraz *et al.*, 2011).

Proteomics research in human diseases has been focused on the development of clinical applications to obtain more precise diagnoses, early detection of diseases and an accurate value of prognostic, and identification of the action mechanism of drugs in pathological cells (which can lead to the characterization of new therapeutic targets). Hence, one of the most relevant clinical applications of proteomics: identification and characterization of specific disease metabolites or metabolites related to prognostic. Because of that, biomarker identification is considered as the last innovation to improve early diagnose, patient monitoring and evaluation of security and efficacy of therapeutic strategies (Dasilva *et al.*, 2012; Díez *et al.*, 2012; Gonzalez-Gonzalez *et al.*, 2012; Matarraz *et al.*, 2011).

Bearing this in mind, it seems proteomics approaches are an ideal methodology to study B-chronic lymphocytic leukaemia for several reasons: (i) little amount of sample is required; (ii) serum profiling of chronic lymphocytic leukaemia (CLL) patients; (iii) capability of sample enrichment; (iv) high throughput that facilitates data analysis in a short time; (v) multiparametric tool which allows the simultaneous study of many parameters.

Here, it is reviewed the proteomics approaches currently implemented for this specific disease having in mind the major disease characteristics which will be described during the chapter.

Epidemiology and incidence

CLL is the most frequent leukaemia in western countries, accounting up to 30% of all leukaemias (Morton *et al.*, 2006). Small lymphocytic lymphoma (SLL), showing lymphoid organ invasion with cells sharing similar features to CLL, is a related entity and they are considered together. It is a clonal lymphoproliferative disorder of morphologically and immunophenotypically mature B lymphocytes. CLL presents in twice the number of males versus females and is most frequent in Caucasian than in African American people (Redaelli *et al.*, 2004). CLL presents the stronger familiar relationship among leukaemias, and it is not unusual to have several members of the same family affected. The genetic factors involved in this behaviour remain unknown.

CLL is the most frequent leukaemia in Western adults, with a prevalence of 20% in people older than 50 years, and its incidence increases with age. CLL is a heterogeneous disease with a variable course. Normally, it presents an indolent course with a 5-year survival rate of 50% and a 10-year survival rate of about 30%, but transformation results in about 5–10% of cases in disease progression, resistance to therapy and very poor prognosis leading to death (Inamdar and Bueso-Ramos, 2007). Actually in many cases, and given the old age of these patients, many of them remained untreated and finally die for reasons not related to the disease.

Clinical presentation

Clinical features and staging

A diagnosis of CLL is based on a persistent lymphocytosis due to the proliferation of clonal B lymphocytes. Traditionally and according to revised guidelines at the 2008 International Workshop on Chronic Lymphocytic Leukaemia, these patients should have more than 5000 B-lymphocytes/μl (Hallek *et al.*, 2008). Although CLL primarily involves peripheral blood and bone marrow, other lymphoid organs, such as lymph nodes and spleen may be involved. Precisely, if this organ involvement does not concur without the lymphocytosis, the term 'small lymphocytic leukaemia' is preferred (Pangalis *et al.*, 1999).

Mild anaemia and thrombocytopenia are not uncommon as a consequence of bone marrow infiltration. Consequently, CLL patients have significantly impaired immune response and prone infection, since these clonal B-cells are not immunologically active.

Diagnosis begins by analysis of peripheral blood. Apart of the lymphocytosis, blood smear show characteristic small lymphocytes with condensed chromatin, 'soccer-ball', and characteristic 'smudge cells', corresponding to degenerated cells. They may have an abnormal high number of pro-lymphocyte cells, larger than CLL cells and with more cytoplasm. Bone marrow aspirates also show lymphocytosis with variable and progressive infiltration. Bone marrow smear is not employed for diagnosis but may be interesting in following the progression of the disease and chemotherapy effects.

Two staging systems are mostly employed to define the disease, the Rai system (Rai *et al.*, 1975), most often used in the United States, and the Binet, used mostly in Europe (Binet *et al.*, 1981). Both systems are very simple and do not require complex laboratory equipment since they are essentially based in leucocyte counting and lymphoid organ involvement. However, they do not consistently predicts patient's clinical outcome and other factors are required for prognosis (Inamdar and Bueso-Ramos, 2007). Adverse factors are diffuse pattern of bone marrow involvement, advanced age, male gender, increasing lymphocyte doubling time (increasing the lymphocytosis) and high serum levels of certain proteins (i.e. beta-2-microglobulin), among others. Genetic lesions and other molecular issues related to prognosis are discussed below.

Finally, another leukaemic transformation may also concurs, typically a high-grade non-Hodgkin lymphoma, most frequently diffuse large B-cell lymphoma (DLBCL). This so-called 'Richter syndrome' occurs in about 5–10% of CLL patients, and it shows an aggressive behaviour with a median survival of 4–5 months (Nakamura and Abe, 2003).

Immunophenotypic characterization

Differential diagnosis of CLL from other B-cell proliferation syndromes requires the phenotypic characterization of peripheral B-cells by flow cytometry (Fig. 4.1). These cells present

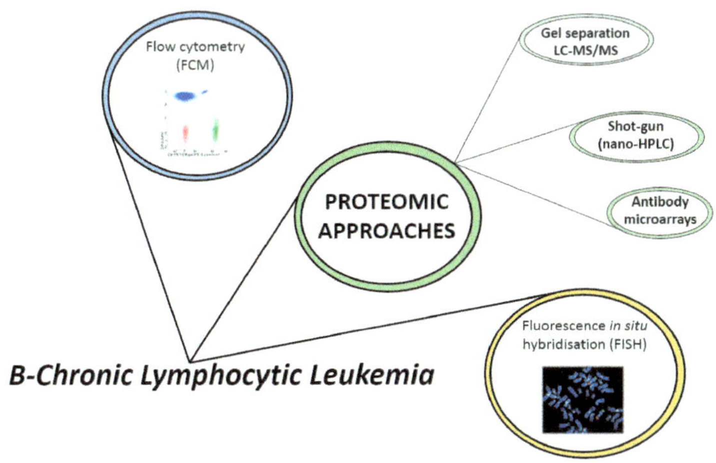

Figure 4.1 Traditional and innovative strategies for B-chronic lymphocytic leukaemia (B-CLL) studies. Traditionally, flow cytometry (FCM) and fluorescence *in situ* hybridization (FISH) have been the techniques of choice for B-CLL analysis. Recently, proteomic approaches, such as antibody microarrays, shot-gun methods and gel separation coupled to LC-MS/MS, have emerged as promising tools.

protein expression with no counterpart to normal cells or any other abnormal status. A distinctive marker is the expression of the CD5 receptor, a common T-cell marker but rarely expressed on normal B-cells. There is a normal B lymphocyte population expressing this marker, the B1 cells, but CLL does not seem to represent its leukaemic counterpart.

There is a characteristic dim expression of immunoglobulin receptors, primarily IgM, and IgD, and less frequently, IgG. A concomitant immunoglobulin light chain restriction κ or λ indicates the clonality of leukaemic populations. A small portion of these patients show oligoclonality, with at least two different B-cell clones. Common B-cell markers such as CD19 and, more interestingly CD20, show low expression, an issue that contributes to differentiate these cells from normal and even other leukaemic B-cells. Other B-cell markers also expressed by these cells includes CD22, CD43, CD79a and CD11c. CD79b, another B-cell marker, is not expressed even at a low level, whereas it is consistently expressed in other B-cell neoplasms. Most mature B-cells and B-cell leukaemia express the FMC7 marker, but again, it is not expressed in most CLL cases, contributing to phenotypic identification.

Molecular features

Chromosomal abnormalities are detected in about 50% of patients by conventional cytogenetic analysis and multiple chromosomal abnormalities are indicators of poor prognosis. Most circulating CLL cells are in a resting (G_0) state. Consequently, conventional cytogenetics methods, valid for the analysis of proliferating cells, have to be implemented to be able to detect many of these abnormalities.

The most common chromosomal abnormality detected by conventional cytogenetics is the 'trisomy 12', in which there is an extra chromosome 12, present in up to 20% of CLL patients. Meanwhile reciprocal translocations are characteristic of other B-cell neoplasms, they are extremely rare in CLL. A 13q deletion is present in about 50% of CLL patients and it is associated to a good prognosis. Other common chromosomal abnormalities are deletions in chromosomes 11q (18%) and 17p (7%), conferring a poor prognosis (Dohner *et al.*, 2000).

Somatic hypermutation (SHM) of genes coding for variable regions of immunoglobulin receptors (IgV) happens in germinal centres of lymphoid organs. CLL cells present a mature B-cell aspect but only about 60% of these clones show mutated genes against 40% of non-mutated immunoglobulin genes. Although not yet fully understood, it would mean that some cells go through the germinal centre T-cell dependent path, whereas other clones would not go into it and are activated by a T-cell independent mechanism (Qiu *et al.*, 2013).

This dichotomy is very interesting in clinical terms since patients with non-mutated clones show a more aggressive disease and worse prognostic compared with patients with mutated genes. However, current mutation analysis is technically demanding and beyond the capacity of many routine laboratories. Consequently, surrogate markers of this mutational status have been developed (Bouley *et al.*, 2006; Tracey *et al.*, 2008). Thus, higher CD38 expression, detected by flow cytometry, has been found to correlate with a non-mutated status and therefore a more aggressive disease. However, several studies did not confirm this predictive value and expression may change in various cases, therefore CD38 plus mutational status is better predictor than any of them alone (Chang and Cleveland, 2002; Hamblin *et al.*, 2002). The expression of the tyrosine kinase *ZAP-70* appears to be the best surrogate marker predicting mutational status in more of 90% of cases in several studies. Zap-70 detection by flow cytometry is technically demanding for being a intracellular protein, but expression remains constant during the disease and still is a much more convenient alternative to CD38 (Crespo *et al.*, 2003). In addition, in many occasions the determination of positivity of Zap-70 requires high-trained technicians because it is based on relative signal compared with Zap-70 in T-cells, which are used as negative reference cells. Aberrant somatic hypermutation (ASHM), a common mechanism of genetic lesions in other B-cell transformations, is absent in CLL (Pasqualucci *et al.*, 2001).

CLL is a clear case of antigen-experienced tumour cells. Analysis of the rearranged IgV genes in leukemic clones of different patients shows certain combinations more frequently present than expected on basis of the normal repertoire of IgV genes, and independently of the mutational status (Chiorazzi and Ferrarini, 2003). Even more surprising is to find conserved Ig rearrangements even at the level of CDR3s of heavy and light chain in unrelated CLL patients, named 'stereotyped receptors' (Stamatopoulos *et al.*, 2007; Murray *et al.*, 2008). This strongly would suggest the key role of certain antigens to promote this leukaemia. The quest for candidate antigens remains open and microbial antigens (bacterial, fungi, etc.) have been implicated, but also auto-antigens (Catera *et al.*, 2008; Lanemo Myhrinder *et al.*, 2008). All in all, it is evident that many questions are still unsolved and it is necessary to elucidate the implicated mechanisms to understand CLL in a better way.

Monoclonal B lymphocytosis (MBL)

Monoclonal B lymphocyte populations, in otherwise healthy people, were already described long time ago in familiar studies of CLL patients but also in non-related people (Han *et al.*, 1984; Kimby *et al.*, 1989). The interest on these cases arose because most of these cells were phenotypically indistinguishable of CLL cancer cells (CLL-like) and other molecular aspects of the disease including IgV mutational status, stereotyped receptors, etc. were also demonstrated (Rawstron, 2009; Shim *et al.*, 2010).

These clonal populations were early identified in healthy individuals with a subtle lymphocytosis (Monoclonal B Lymphocytosis, MBL) approaching higher levels characteristic of CLL patients. Guidelines for appropriate CLL diagnosis were set by an International Workshop on CLL (IWCLL) in 1989, defining CLL lymphocytosis as $> 5 \times 10^9$ lymphocytes/l. Further re-definition by the IWCLL in 2008 propose absolute B-lymphocyte counting (and not total lymphocyte) as diagnostic criteria, and now CLL is diagnosed considering a value of $> 5 \times 10^9$ B-lymphocytes/l (Hallek *et al.*, 2008). This change even led to a reclassification of many people, previously considered, patients, as 'healthy' individuals (Shanafelt *et al.*, 2008).

Prevalence of MBL in the adult population may even raise to 14% according to a systematic review in MBL population studies (Shim *et al.*, 2010).

Genetic and epigenetic analysis of CLL

CLL is a B-cell malignancy with a strong familial component; risk is increased from 7- to 8-fold in first-degree relatives (Goldin *et al.*, 2009), but to date inherited genetic reasons for the disease remain unknown. There is no evidence of major disease-candidate genes and a model of multiple low-risk variants is most likely.

Next-generation sequencing (NGS) techniques are helping to uncover the genetic complexity and heterogeneity of CLL (reviewed by Rodriguez-Vicente *et al.*, 2013). The most common mutated genes are *NOTCH1* (~12% of CLL patients) and SF3B1 (~10%) (Puente *et al.*, 2011; Quesada *et al.*, 2011). *NOTCH1* is a gene involved in signalling, mutated in more than 50% of T-ALL patients (a lymphoblastic leukaemia) and most of these CLL patients carry unmutated IGHV genes and are associated with poor prognosis (Puente *et al.*, 2011). SF3B1 is a protein involved in the splicing machinery and CLL mutations are also related to poor prognosis (Quesada *et al.*, 2011).

Genome-wide associations studies (GWAS) tagging single nucleotide polymorphisms have identified more than 13 loci associated to the disease (reviewed by Martin-Subero *et al.*, 2013), several of them are strong candidates for participating in the disease. One of these loci refers to the *IRF-4* gene, a transcription factor involved in key regulation steps of B-cell activation and maturation (Di Bernardo *et al.*, 2008). Other study have found association of a group of genes intervening in apoptosis, including *FAS*, *BCL-2* and *NOXA*, previously implicated in the survival of chronic lymphocytic leukaemia cells (Berndt *et al.*, 2013).

In addition, CLL epigenome has been explored in the last years (reviewed by Martin-Subero *et al.*, 2013). As in other cases of leukaemia, CLL DNA is hypomethylated when compared to normal peripheral blood mononuclear cells (Kulis *et al.*, 2012). Otherwise, specific gene silencing by promoter hypermethylation of suppressor genes have been also described (Rahmatpanah *et al.*, 2009). Other genes are also epigenetically silenced and even Zap-70 protein levels correlate with specific CpG islands methylation within the gene (Martin-Subero *et al.*, 2013).

Proteomics profiles in B-cell malignancies

The human genome consists of approximately 30,000 genes, which with alternate splicing, sequence deletion and post-translational modifications can be translated into more than 10^6 distinct protein species. The exact number of proteins will depend on cell type, metabolic or disease status, but signalling proteins or those acting at key signalling or metabolic control points will be in low abundance, while housekeeping/constitutive, metabolic enzymes and cytoskeletal proteins will be often in very high abundance. Protein levels are also highly dynamic and subject to rapid changes, according to the physiological status of the cell.

B-cell malignancies are more amenable to proteomic analyses as white blood cells can be obtained from patients relatively easily, affinity purified and in sufficient quantity for meaningful analysis.

This, however, does carry the assumption that the proteome of peripheral malignant B-cells reflects their point of origin in the germinal centre within the lymph node or bone marrow environments. For many proteins this may be true but several proteins may show significant differences. Thus, for example, B-CLL cells from lymph nodes express higher Zap-70 levels than B-CLL cells from peripheral blood (Butrym *et al.*, 2013). However, Noxa, a pro-apoptotic protein, was up-regulated at both the mRNA and protein levels in peripheral B-CLLs compared with lymph node cells. Therefore, if possible, proteomic studies obtained from peripheral cells need to be compared to lymph node analyses.

A variety of proteomics approaches have been used to investigate B-cell malignancies (Fig. 4.1), including protein antibody microarrays, 2-D-gel electrophoresis coupled to MALDI-TOF, 'shot-gun proteomics' using nano-HPLC or gel separations coupled to LC-MS/MS.

Protein antibody microarray studies offer high sensitivity and throughput but are limited to the availability of high quality antibodies and cannot identify unknown proteins, but potentially it should be possible to tailor antibody arrays to selectively identify a particular disease or a set of markers which could be used in prognostic decision and/or therapeutic rationales.

In this chapter, we have minimized descriptions of available methodologies and focused on biological and clinical significance of the studies in B-cell malignancies.

In addition, the success of proteomics studies on B-cell malignancies needs to be measured in terms of outcomes, such as identifying proteins that (a) contribute to our understanding of B-cell malignancies; (b) could be used for diagnosis or prognosis; (c) are potential therapeutic targets.

Proteome analysis of B-cell maturation

B-cell development is leaded to establish a diverse of peripheral B-cells that recognize pathogens while maintaining tolerance to itself.

Activated B-cells proliferate and mature under the influence of various cell types and cytokines, and ultimately differentiate into antibody-secreting plasma or memory cells, the final mediators of the humoral immune response.

Maturing B-cells proceed through several distinct developmental stages, during which VDJ rearrangements take place within antibody-coding genes. Cells at each developmental stage can be distinguished by the expression of specific cells surface proteins, as well as by the extent of the rearrangement of immunoglobulin genes. Mutations in genes involved in the production of components of the immune response result in defective or aberrant B-cell function and development causing immunodeficiency diseases.

B-cell differentiation is triggered by the crosslinking of the B-cell receptor (BCR), which is one of the most studied molecules that transmit messages from outside the cell into the cell. The BCR structure and signalling pathways have been extensively described. The BCR consists of the antigen-binding membrane-bound immunoglobulin (mIg) in association with accessory signal transduction components Igα (CD79a) and Igβ (CD79b). As a consequence of BCR assembled, a dynamic network of intracellular signalling cascades leads to the activation of nuclear transcription factors and regulation of gene expression.

Signals are transmitted by means of protein phosphorylation, modifications and interaction. Many enzymes, such as protein tyrosine kinases, phosphatases and phospholipases, are major components of these networks. The first signals are activation of the Src-family Fyn, Lyn and Blk as well as Syk and Bruton tyrosine kinase (Btk).

Together with adapter proteins, PTKs are coupled to downstream signalling cascades, including the mobilization of Ca^{2+} ions, the activation of Ras and the protein kinase C-pathway and the re-organization of cytoskeleton architecture. Many factors, such as the developmental stage of B-cell and co-receptor signalling, influence the BCR response. Immature B-cells respond to BCR cross-linking by increasing the expression of genes involved in receptor editing or apoptosis. The molecular mechanism by which the BCR can elicit differential maturation state-specific responses is not fully understood. The differential functional responses of BCR-coupled signals depend upon several parameters, such as signal strength and duration, subcellular localization of the signal, maturation-restricted expression of downstream effectors and modulation of the signal by possible co-receptors. The engagement of mIgM leads to B-cell proliferation *in vitro*.

Salonen *et al.* (2006) studied several non-quantitative proteomics approaches the change in protein abundance involved in B-cell maturation. In this study, the identification of proteome profiles of stimulated and non-stimulated Ramos B-cells were performed. Anti-IgM stimulation of the cells resulted in significant differences ($\geq$ 2-fold) in the protein abundance close to 100 proteins and differences in post-translational protein modifications (Salonen *et al.*, 2006).

At least, 48 up- and down-regulated proteins were identified by mass spectrometric (MS) methods and database searches. A few of these proteins are directly related to the functional and morphological characteristics of B-cells, such as cytoskeleton rearrangement and intracellular signalling triggered formation of BCR. Overall, these results provide deeper insights into the process of B-cell maturation and may lead to novel therapeutic strategies for immunodeficiency.

Quantitative proteomics; 'shot gun' proteomic studies on B-cell malignancies

'Shot-gun' proteomics has emerged as a powerful technique for analysing membrane proteomes because hydrophobic membrane and basic proteins are difficult to resolve with 2-DE and an alternative approach to analyse membrane proteins is to use 1-D SDS-PAGE and MS. ¨Shot-gun¨ proteomics basically exploits the power of modern LC-MS/MS tandem mass spectrometers to discriminate between thousands of peptides, which can be individually separated and then sequenced by fragmentation using collision induced dissociation (CID). In combination with available protein databases and sophisticated bioinformatics techniques it is now possible to identify many different proteins in one sample. Commonly, two approaches are used:

1 *Multidimensional LC-MS/MS*. The protein mixture is digested using proteases (i.e. trypsin) and then the peptides are separated by cation exchange chromatography followed by reverse phase chromatography to yield the signature peptides which are identified in the tandem mass spectrometers.
2 *Gel-based shotgun proteomics*. The proteins are separated by molecular weight on 1-D SDS-PAGE gels which are sequentially sliced and subjected to in-gel trypsinolysis to yield the peptides which are identified by LC-MS/MS.

Both shotgun approaches are equally efficient at identifying large numbers of proteins, and the only major difference between the two approaches is that the gel-based approach gives extra information on the protein, in that detection of the protein with an anomalous molecular weight can be indicative of proteolytic cleavage or degradation (e.g. during caspase-dependent cell death) or PTM (i.e. glycosylation).

Shotgun proteomics is a powerful tool and coupled with appropriate quantitative strategies can deliver important information on protein changes in B-cell malignancies. A great number of methodologies have been developed to provide quantitative data. These techniques involve pre- or post-labelling of proteins with stable isotope tags, which can be detected and quantified by MS.

Stable isotope labelling of amino acids in culture (SILAC) is a relatively non-invasive technique in which cells are pre-labelled in media containing appropriate ^{13}C- and/or

^{15}N-labelled amino acids. Vogler *et al.* (2009) studied, by SILAC approaches, the resistance mechanism of ABT-263 drug in CLL cells within lymph nodes and in peripheral blood. The study revealed that this resistance occurred upstream of mitochondrial perturbation and involved *de novo* synthesis of the anti-apoptotic proteins Bcl-X (L) and Bcl2A1. CLL cells within lymph nodes are more resistant to apoptosis than those in peripheral blood (Vogler *et al.*, 2009).

Another useful approach for primary leukemic cells is to post-label the protein with ICAT or the peptides using iTRAQ (isobaric tagging for relative quantification). The iTRAQ technique (Fig. 4.2) employs four or eight isobaric reagents (each isobaric reagent consists of reporter, balance and reactive groups to give the same mass), TAG peptides which are then identified by MS/MS. The reactive group attaches the tag to N-terminal amino moieties and lysine with reporter groups (masses of 114, 115, 116 and 117 Da) and complementary balance groups (masses of 31, 30, 29 and 28 Da). The compensating masses of reporter and balance groups have the same mass and a particular peptide, tagged by any of the iTRAQ reagents, has the same mass to charge ratio in the MS spectrum. As both control and test samples are combine, this increases the sensitivity of peptide detection and during MS/MS, fragmentation releases a distinctive reporter ion that can be used for relative quantification of the peptide.

Post-labelling with ICAT or iTRAQ can be used with primary leukemic cells, and cICAT (cleavable for of ICAT) has been used to analyse mutated-CLL (M-CLL) and unmutated-CLL (UM-CLL). Barnidge *et al.* (2005) identified by iTRAQ several different proteins (i.e. cytochrome c oxidase subunit, COX, etc.) up-regulated in 6 M-CLL patients.

One additional label-free method has been developed based on the number of peptides (protein abundance index, PAI) or spectra [spectral abundance factor (SAF)]. In both cases, it is possible to obtain data regarding the relative abundance for a given protein in each sample. In addition, selective reaction monitoring (SRM) measurements of a limited set of internal reference standards were used to determine the absolute protein concentrations of more than thousands of proteins. Although these methods are not perfect for absolute quantification, at least by these methods it is possible to identify potential changes in normal and malignant cells, which can be validated by using Western-blot (WB) RT-PCR and/or immune-histochemistry. Boyd *et al.* (2009) in a recent study of MCL, identified several proteins such as CD20, CD79b, CD22, CD31, CD11a, CD50, CD82, and CD44 and validated them by WB and RT-PCR.

Conclusions and future directions

Recently, proteomics technology has made huge progress thanks to the power of the sensitive mass spectrometers, databases and bioinformatics software, which provide deep knowledge of the protein changes that may underlie many diseases. However, the use of many of these proteomics approaches remains a challenging problem.

To overcome the limited protein coverage of current proteomic strategies, a more target approach could be used to increase the discovery rate, by fractionating the cell into component fractions, such as nuclei, organelle, cytoplasmic proteins, etc. In addition, it will be relevant to applied target proteomics approaches on some important B-cell signalling complexes, which could influence the response of the malignant B-cells to therapeutic agents.

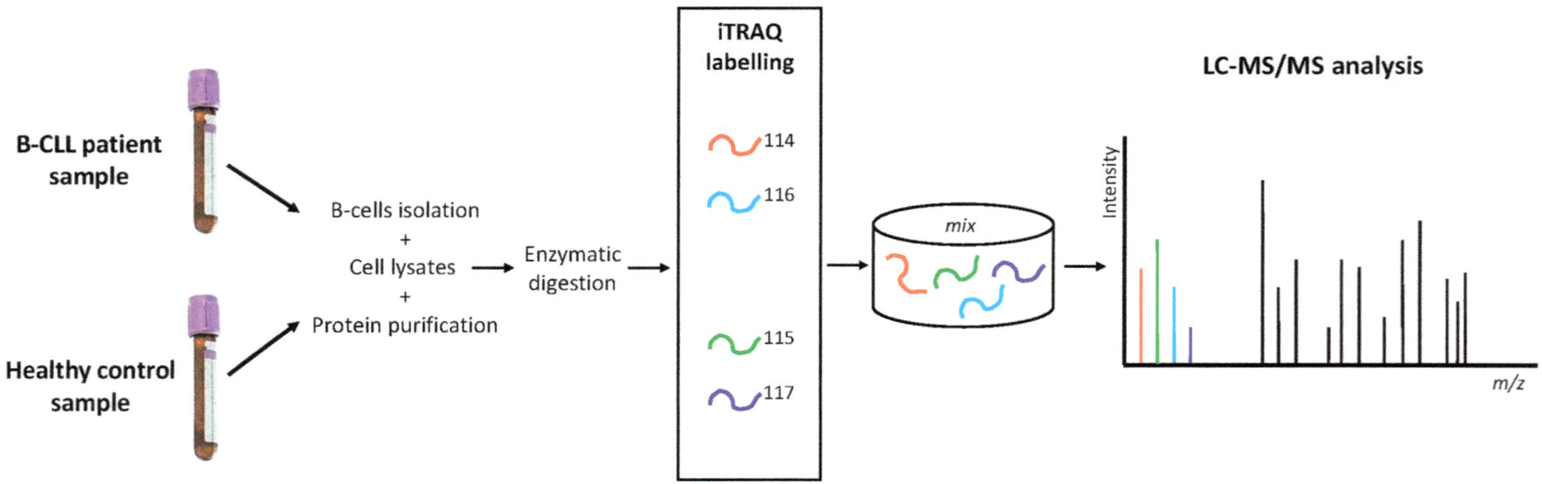

Figure 4.2 Workflow of iTRAQ processing. Proteins from control and test samples are labelled with different isobaric tags and then combined for LC-MS/MS analysis. MS spectrum is employed for relative quantification of peptides.

The success of any proteomics approach might be only measured in terms of the outcomes: (i) help to understand the disease; (ii) identification of biomarkers useful in diagnosis or prognosis; (iii) identification of proteins as novel targets for new drugs. In addition, any target proteomics approach on B-cell malignancies might be integrated and validated with biological and clinical studies.

References

Barnidge, D.R., Jelinek, D.F., Muddiman, D.C., and Kay, N.E. (2005). Quantitative protein expression analysis of CLL B-cells from mutated and unmutated IgV(H) subgroups using acid-cleavable isotope-coded affinity tag reagents. J. Proteome Res. *4*, 1310–1317.

Berndt, S.I., Skibola, C.F., Joseph, V., Camp, N.J., Nieters, A., Wang, Z., Cozen, W., Monnereau, A., Wang, S.S., Kelly, R.S., *et al.* (2013). Genome-wide association study identifies multiple risk loci for chronic lymphocytic leukemia. Nat. Genet. *45*, 868–876.

Binet, J.L., Auquier, A., Dighiero, G., Chastang, C., Piguet, H., Goasguen, J., Vaugier, G., Potron, G., Colona, P., Oberling, F., *et al.* (1981). A new prognostic classification of chronic lymphocytic leukemia derived from a multivariate survival analysis. Cancer *48*, 198–206.

Bouley, J., Deriano, L., Delic, J., and Merle-Beral, H. (2006). New molecular markers in resistant B-CLL. Leuk. Lymphoma *47*, 791–801.

Butrym, A., Majewski, M., Dzietczenia, J., Kuliczkowski, K., and Mazur, G. (2013). High CD74 expression correlates with ZAP70 expression in B-cell chronic lymphocytic leukemia patients. Med. Oncol. *30*, 560

Catera, R., Silverman, G.J., Hatzi, K., Seiler, T., Didier, S., Zhang, L., Herve, M., Meffre, E., Oscier, D.G., Vlassara, H., *et al.* (2008). Chronic lymphocytic leukemia cells recognize conserved epitopes associated with apoptosis and oxidation. Mol. Med. *14*, 665–674.

Chang, C.C., and Cleveland, R.P. (2002). Conversion of CD38 and/or myeloid-associated marker expression status during the course of B-CLL: association with a change to an aggressive clinical course. Blood *100*, 1106.

Chiorazzi, N., and Ferrarini, M. (2003). B-cell chronic lymphocytic leukemia: lessons learned from studies of the B-cell antigen receptor. Annu. Rev. Immunol. *21*, 841–894.

Crespo, M., Bosch, F., Villamor, N., Bellosillo, B., Colomer, D., Rozman, M., Marce, S., Lopez-Guillermo, A., Campo, E., and Montserrat, E. (2003). ZAP-70 expression as a surrogate for immunoglobulin-variable-region mutations in chronic lymphocytic leukemia. N. Engl. J. Med. *348*, 1764–1775.

Dasilva, N., Díez, P., Matarraz, S., Gonzalez-Gonzalez, M., Paradinas, S., Orfao, A., and Fuentes, M. (2012). Biomarker discovery by novel sensors based on nanoproteomics approaches. Sensors (Basel). *12*, 2284–2308.

Di Bernardo, M.C., Crowther-Swanepoel, D., Broderick, P., Webb, E., Sellick, G., Wild, R., Sullivan, K., Vijayakrishnan, J., Wang, Y., Pittman, A.M., *et al.* (2008). A genome-wide association study identifies six susceptibility loci for chronic lymphocytic leukemia. Nat. Genet. *40*, 1204–1210.

Díez, P., Dasilva, N., González-González, M., Matarrraz, S., Casado-Vela, J., Orfao, A., and Fuentes, M. (2012). Data analysis strategies for protein microarrays. Microarrays. *1*, 64–83.

Dohner, H., Stilgenbauer, S., Benner, A., Leupolt, E., Krober, A., Bullinger, L., Dohner, K., Bentz, M., and Lichter, P. (2000). Genomic aberrations and survival in chronic lymphocytic leukemia. N. Engl. J. Med. *343*, 1910–1916.

Goldin, L.R., Bjorkholm, M., Kristinsson, S.Y., Turesson, I., and Landgren, O. (2009). Elevated risk of chronic lymphocytic leukemia and other indolent non-Hodgkin's lymphomas among relatives of patients with chronic lymphocytic leukemia. Haematologica *94*, 647–653.

Gonzalez-Gonzalez, M., Jara-Acevedo, R., Matarraz, S., Jara-Acevedo, M., Paradinas, S., Sayagues, J.M., Orfao, A., and Fuentes, M. (2012). Nanotechniques in proteomics: protein microarrays and novel detection platforms. Eur. J. Pharm. Sci. *45*, 499–506.

Hallek, M., Cheson, B.D., Catovsky, D., Caligaris-Cappio, F., Dighiero, G., Dohner, H., Hillmen, P., Keating, M.J., Montserrat, E., Rai, K.R., *et al.* (2008). Guidelines for the diagnosis and treatment of chronic lymphocytic leukemia: a report from the International Workshop on Chronic Lymphocytic Leukemia updating the National Cancer Institute-Working Group 1996 guidelines. Blood *111*, 5446–5456.

Hamblin, T.J., Orchard, J.A., Ibbotson, R.E., Davis, Z., Thomas, P.W., Stevenson, F.K., and Oscier, D.G. (2002). CD38 expression and immunoglobulin variable region mutations are independent prognostic variables in chronic lymphocytic leukemia, but CD38 expression may vary during the course of the disease. Blood *99*, 1023–1029.

Han, T., Ozer, H., Gavigan, M., Gajera, R., Minowada, J., Bloom, M.L., Sadamori, N., Sandberg, A.A., Gomez, G.A., and Henderson, E.S. (1984). Benign monoclonal B-cell lymphocytosis – a benign variant of CLL: clinical, immunologic, phenotypic, and cytogenetic studies in 20 patients. Blood *64*, 244–252.

Inamdar, K.V., and Bueso-Ramos, C.E. (2007). Pathology of chronic lymphocytic leukemia: an update. Ann. Diagn. Pathol. *11*, 363–389.

Kimby, E., Mellstedt, H., Bjorkholm, M., and Holm, G. (1989). Clonal cell surface structures related to differentiation, activation and homing in B-cell chronic lymphocytic leukemia and monoclonal lymphocytosis of undetermined significance. Eur. J. Haematol. *43*, 452–459.

Kulis, M., Heath, S., Bibikova, M., Queiros, A.C., Navarro, A., Clot, G., Martinez-Trillos, A., Castellano, G., Brun-Heath, I., Pinyol, M., *et al.* (2012). Epigenomic analysis detects widespread gene-body DNA hypomethylation in chronic lymphocytic leukemia. Nat. Genet. *44*, 1236–1242.

Lanemo Myhrinder, A., Hellqvist, E., Sidorova, E., Soderberg, A., Baxendale, H., Dahle, C., Willander, K., Tobin, G., Backman, E., Soderberg, O., *et al.* (2008). A new perspective: molecular motifs on oxidized LDL, apoptotic cells, and bacteria are targets for chronic lymphocytic leukemia antibodies. Blood *111*, 3838–3848.

Martin-Subero, J.I., Lopez-Otin, C., and Campo, E. (2013). Genetic and epigenetic basis of chronic lymphocytic leukemia. Curr. Opin. Hematol. *20*, 362–368.

Matarraz, S., Gonzalez-Gonzalez, M., Jara, M., Orfao, A., and Fuentes, M. (2011). New technologies in cancer. Protein microarrays for biomarker discovery. Clin. Transl. Oncol. *13*, 156–161.

Morton, L.M., Wang, S.S., Devesa, S.S., Hartge, P., Weisenburger, D.D., and Linet, M.S. (2006). Lymphoma incidence patterns by WHO subtype in the United States, 1992–2001. Blood *107*, 265–276.

Murray, F., Darzentas, N., Hadzidimitriou, A., Tobin, G., Boudjogra, M., Scielzo, C., Laoutaris, N., Karlsson, K., Baran-Marzsak, F., Tsaftaris, A., *et al.* (2008). Stereotyped patterns of somatic hypermutation in subsets of patients with chronic lymphocytic leukemia: implications for the role of antigen selection in leukemogenesis. Blood *111*, 1524–1533.

Nakamura, N., and Abe, M. (2003). Richter syndrome in B-cell chronic lymphocytic leukemia. Pathol. Int. *53*, 195–203.

Pangalis, G.A., Angelopoulou, M.K., Vassilakopoulos, T.P., Siakantaris, M.P., and Kittas, C. (1999). B-chronic lymphocytic leukemia, small lymphocytic lymphoma, and lymphoplasmacytic lymphoma, including Waldenstrom's macroglobulinemia: a clinical, morphologic, and biologic spectrum of similar disorders. Semin. Hematol. *36*, 104–114.

Pasqualucci, L., Neumeister, P., Goossens, T., Nanjangud, G., Chaganti, R.S., Kuppers, R., and Dalla-Favera, R. (2001). Hypermutation of multiple proto-oncogenes in B-cell diffuse large-cell lymphomas. Nature *412*, 341–346.

Puente, X.S., Pinyol, M., Quesada, V., Conde, L., Ordonez, G.R., Villamor, N., Escaramis, G., Jares, P., Bea, S., Gonzalez-Diaz, M., *et al.* (2011). Whole-genome sequencing identifies recurrent mutations in chronic lymphocytic leukaemia. Nature *475*, 101–105.

Qiu, X., Sun, X., He, Z., Huang, J., Hu, F., Chen, L., Lin, P., You, M.J., Medeiros, L.J., and Yin, C.C. (2013). Immunoglobulin gamma heavy chain gene with somatic hypermutation is frequently expressed in acute myeloid leukemia. Leukemia *27*, 92–99.

Quesada, V., Conde, L., Villamor, N., Ordonez, G.R., Jares, P., Bassaganyas, L., Ramsay, A.J., Bea, S., Pinyol, M., Martinez-Trillos, A., *et al.* (2011). Exome sequencing identifies recurrent mutations of the splicing factor SF3B1 gene in chronic lymphocytic leukemia. Nat. Genet. *44*, 47–52.

Rahmatpanah, F.B., Carstens, S., Hooshmand, S.I., Welsh, E.C., Sjahputera, O., Taylor, K.H., Bennett, L.B., Shi, H., Davis, J.W., Arthur, G.L., *et al.* (2009). Large-scale analysis of DNA methylation in chronic lymphocytic leukemia. Epigenomics *1*, 39–61.

Rawstron, A.C. (2009). Monoclonal B-cell lymphocytosis. Hematol./Educ. Progr. Am. Soc. Hematol. Soc. Hematol. Progr. 430–439.

Redaelli, A., Laskin, B.L., Stephens, J.M., Botteman, M.F., and Pashos, C.L. (2004). The clinical and epidemiological burden of chronic lymphocytic leukaemia. Eur. J. Cancer Care (Engl). *13*, 279–287.

Salonen, J.M., Valmu, L., Ronnholm, G., Kalkkinen, N., and Vihinen, M. (2006). Proteome analysis of B-cell maturation. Proteomics *6*, 5152–5168.

Shanafelt, T.D., Kay, N.E., Call, T.G., Zent, C.S., Jelinek, D.F., LaPlant, B., Morice, W.G., and Hanson, C.A. (2008). MBL or CLL: which classification best categorizes the clinical course of patients with an absolute lymphocyte count $\geq 5 \times 10^9\ L^{-1}$ but a B-cell lymphocyte count $<5 \times 10^9\ L^{-1}$? Leuk. Res. *32*, 1458–1461.

Shim, Y.K., Middleton, D.C., Caporaso, N.E., Rachel, J.M., Landgren, O., Abbasi, F., Raveche, E.S., Rawstron, A.C., Orfao, A., Marti, G.E., *et al.* (2010). Prevalence of monoclonal B-cell lymphocytosis: a systematic review. Cytom. B, Clin. Cytom. *78 Suppl 1*, S10–8.

Stamatopoulos, K., Belessi, C., Moreno, C., Boudjograh, M., Guida, G., Smilevska, T., Belhoul, L., Stella, S., Stavroyianni, N., Crespo, M., *et al.* (2007). Over 20% of patients with chronic lymphocytic leukemia carry stereotyped receptors: Pathogenetic implications and clinical correlations. Blood *109*, 259–270.

Tracey, L., Aggarwal, M., Garcia-Cosio, M., Villuendas, R., Algara, P., Sanchez-Beato, M., Sanchez-Aguilera, A., Garcia, J.F., Rodriguez, A., Camacho, F.I., *et al.* (2008). Somatic hypermutation signature in B-cell low-grade lymphomas. Haematologica 93, 1186–1194.

Vogler, M., Butterworth, M., Majid, A., Walewska, R.J., Sun, X.M., Dyer, M.J., and Cohen, G.M. (2009). Concurrent up-regulation of BCL-XL and BCL2A1 induces approximately 1000-fold resistance to ABT-737 in chronic lymphocytic leukemia. Blood *113*, 4403–4413.

Protein Microarrays: A Versatile Tool for Scientific Discovery

5

Johnathan Neiswinger, Jiang Qian, Jin Zhang and Heng Zhu

Abstract

Over the last quarter-century, protein microarray technology has emerged as a prominent field in scientific study. The versatility of the platform, coupled with the ability to characterize thousands of proteins in a parallel and high-throughput manner, has resulted in great strides in our knowledge database. Microarrays consist of three classes, analytical, functional, and reverse-phase, all of which have their own place in biomedical research. Much progress in this technology has been made over the last decade as new methods, particularly in microarray fabrication, have been developed. This chapter will give an overview of new and founding technologies, complete with a series of examples from both the basic and clinical research fields, including microarray fabrication, detection methods (label dependent and independent), binding interactions, post-translational modifications (PTMs), host–microbe interactions, and cancer research.

Introduction

History

In 1983, Tse-Wan Chang spotted a variety of antibodies onto a glass slide in a matrix formation in order to study 'the potential of simultaneous multiple determinations of specific cell surface antigens in one reaction incubation (Chang, 1983).' This study represents one of the first concepts of what we today refer to as a microarray. This notion blossomed into ambient analyte immunoassays, introduced by Roger Ekins in 1989, in which two fluorescently labelled antibodies were used to detect the occupation of a 'sensing' antibody spotted on a slide (Ekins, 1989). Over the next ten years, the concept was transformed into DNA microarrays, which allow for the detection of mRNA expression levels of thousands of genes in parallel.

However, it was found that mRNA levels do not always correlate with protein expression, so the technology expanded further through the creation of protein microarrays in order to overcome some of these limitations (Gygi *et al.*, 1999; Zhu *et al.*, 2001; Kopf *et al.*, 2007). Although DNA microarrays provide a useful platform for biological studies, they cannot emulate the function of what many consider the workhorses of the cell – proteins.

Protein array technology was developed in the early stages by organizing bacterial strains of a cDNA library on a nylon membrane, lysing the bacteria, and immobilizing the total protein complement (Bussow *et al.*, 1998). Although the technology was promising, it suffered from four critical issues. First, the human proteins were expressed in bacteria, so they lacked the critical post-translational modifications found in eukaryotes; second, the bacterial lysis was performed in denaturing conditions, resulting in misfolded proteins; third, the proteins of interest were not separated from the thousands of bacterial proteins from the cell lysate; and finally, the overall density of the array was very low. The purity and density problems were quickly resolved by a variety of groups, who used either purified proteins (MacBeath *et al.*, 2000; Zhu *et al.*, 2000; Ge, 2000) or antibodies (Schweitzer *et al.*, 2002) for their arrays.

In 2001, the first study using these high-density arrays was presented by the Snyder group at Yale (Zhu *et al.*, 2001). This 'proteome' microarray began by expressing approximately 5800 full-length yeast open reading frames (ORFs) individually in yeast, where they were purified as N-terminal GST-fusion proteins. The proteins were then spotted on a glass slide in duplicate robotically to form the first high-density 'proteome' microarray, as it contained more than 75% of the known yeast proteome. Since this time, many more proteome microarrays have been created from viruses, bacteria, plants, and humans (Zhu *et al.*, 2006; Chen *et al.*, 2008; Popescu *et al.*, 2007; Jeong *et al.*, 2012).

Types of protein microarrays

There are three accepted types of protein microarrays that have been created, all with different functions and benefits: analytical, functional, and reverse-phase. In the next section we will define and compare the differences between these three classes of microarrays (Fig. 5.1) and list a number of these arrays that are commercially available (Table 5.1).

Analytical protein microarrays

Analytical microarrays focus on protein detection from a complex mixture of proteins (Fig. 5.1A). The canonical example of this type of array is the antibody array, which uses an assortment of spotted antibodies to capture proteins from a cell lysate. This is most commonly performed using a process called direct protein labelling. In this technique, protein samples are labelled with a fluorophore before binding to the array. Protein labelling can be useful for high-abundance proteins, but suffers from low sensitivity and specificity in target labelling. A sandwich approach was developed to increase the sensitivity by using two separate antibodies to detect the target protein. The 'capture' antibody is immobilized onto the solid phase, where it binds the protein of interest. The second 'reporter' antibody contains the fluorescent label and binds to the target protein, leading to much greater sensitivity and specificity, even down to femtomolar concentrations (Poetz *et al.*, 2005).

Although antibodies are the primary means of protein capture for analytical microarrays, they do have some sizeable disadvantages. For starters, the specificity of the antibodies can vary considerably and can cross-react with many other proteins, resulting in a large amount of false positives. Another hurdle is due to the difficulty in producing large quantities of antibodies in a high-throughput fashion. The advent of recombinant antibodies holds a promising means of overcoming this, but complex production issues need to be addressed before this technology becomes feasible in the research community (Brichta *et al.*, 2005).

Functional protein microarrays

The second class of protein microarrays is the functional protein microarray (Fig. 5.1B). These arrays consist of individually purified proteins immobilized to a glass slide and are used in a variety of biochemical reactions or assays. Unlike the analytical arrays, which focus on detections from a complex protein mixture, functional microarrays generally consist of

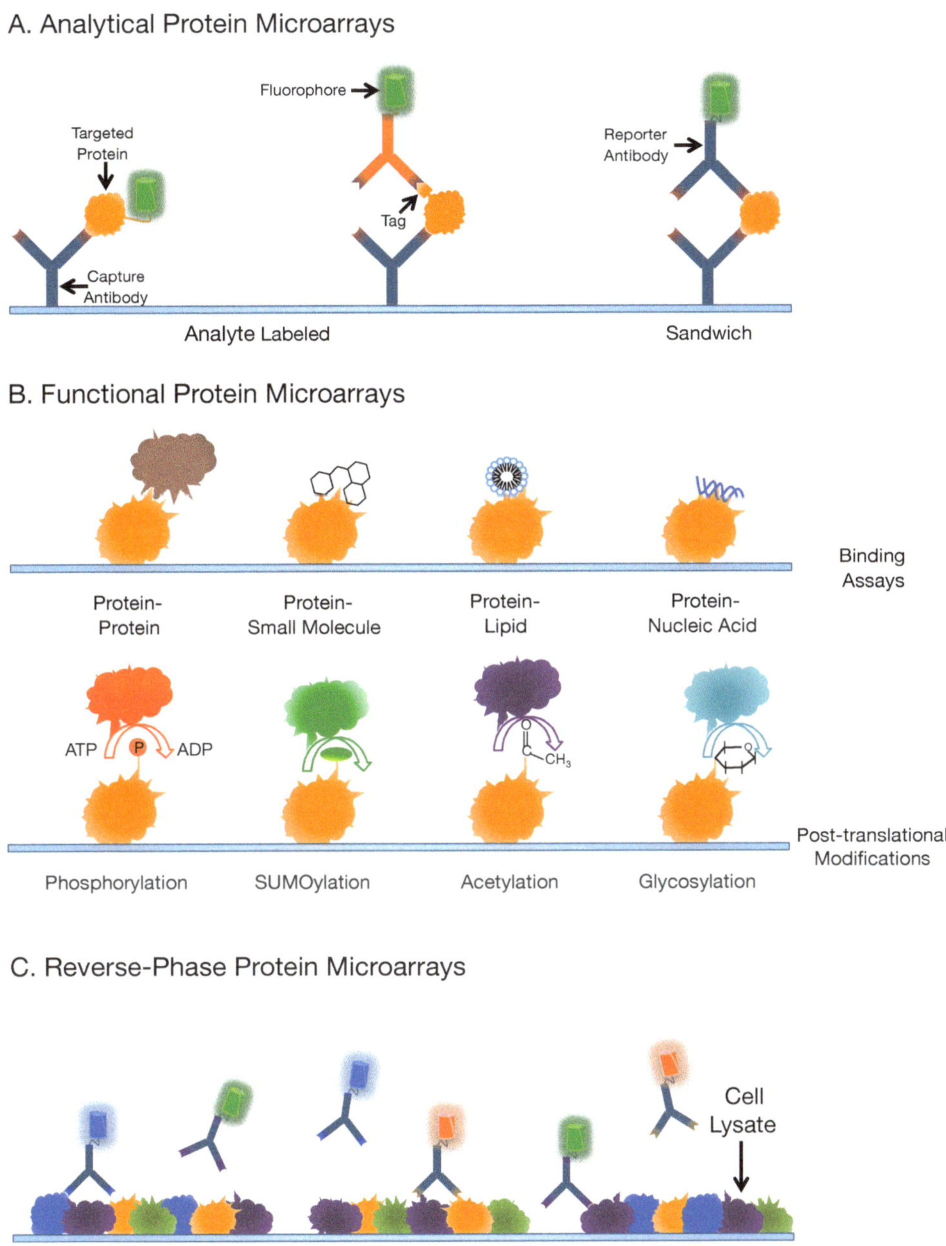

Figure 5.1 The three types of protein microarrays. (A) Analytical protein microarrays generally are fabricated with antibodies on a glass surface. Analyte can be labelled directly, identified via a tag, or identified with a separate labelled antibody. (B) Functional protein microarrays constitute the largest diversity in their applications. Simple binding assays can be performed and are usually tethered to a fluorescent protein, but can also utilize tight non-covalent interactions (i.e. biotin–streptavidin). Post-translational modification studies utilize an enzyme of interest with a labelled cofactor for detections of *de novo* modifications. (C) Reverse-phase protein microarrays consist of probing complex mixtures of proteins, such as a cell lysate, with very specific antibodies. Multicolour assays are possible using this technique, and these are commonly used in biomarker identification studies.

Table 5.1 Commercially available protein microarrays

Product type	Product name	Company	Array type	Protein content
Protein identification antibody arrays	RayBioHuman RTK Phosphorylation Antibody Array	RayBiotech	Analytical	Antibodies against 71 human kinases
	PlasmaScan 380 Antibody Microarray	Arrayit	Analytical	Antibodies for detection of human plasma
	Cytokine Antibody Microarray	Full Moon BioSystems	Analytical	Antibodies against 77 human cytokines
	Kinase Antibody Microarray	Full Moon BioSystems	Analytical	Antibodies against 276 human kinases
Pathway detection antibody arrays	MAPK Pathway Phospho Antibody Array	BioCat	Analytical	185 phospho-protein antibodies for proteins in the MAPK pathways
	Signalling Explorer Antibody Microarray	Full Moon BioSystems	Analytical	1358 antibodies covering multiple pathways
	Wnt Signalling Phospho Antibody Microarray	Full Moon BioSystems	Analytical	227 phospho-antibodies from cell growth, movement, and development pathways
Human protein arrays	ProtoArray	Invitrogen	Functional	9000 human proteins
	Kinex	Kinexus Bioinformatics	Functional	200 human kinases
Pathogen arrays	Arrayit Pathogen Antigen Microarray	Arrayit	Functional	Essential proteins of various pathogens
Cell lysate arrays	SomaPlex	Protein Biotechnologies	Reverse-Phase	Variety of human cancer cell lysates

treating the array with a single purified protein in order to study its interaction with the proteins on the chip. These interactions can vary from binding activities (protein–protein, protein–DNA, protein–lipid, protein–drug) to enzyme–substrate relationships, which are analysed from a biochemical assay performed on the array (phosphorylation, ubiquitylation, acetylation, glycosylation, SUMOylation, nitrosylation) (Poetz *et al.*, 2005; Chen *et al.*, 2006). One large advantage is derived from the sheer amount of diverse experiments that can be applied to the array, but caution must be applied when labelling hits as true *in vivo* interactions. Although many assays involve the use of purified proteins, the interactions may not reflect what actually occurs in a living system. Conversely, some true interactions may not be seen due to the absence of cofactors in the reaction mix. Because of this, any interactions identified on a functional array must be further characterized before conclusions can be drawn.

Reverse-phase protein microarrays

Reverse-phase protein microarrays are unique in that they employ the opposite format to the classical microarray (Fig. 5.1C). Tissues, cell lysates, or cell fractionations from different states are immobilized to a glass slide, where they can be treated with a variety of probes to identify a protein of interest. Cloud Paweletz was one of the first to use this method in his analysis of histological changes in prostate cancer patients (Paweletz *et al.*, 2001). In this study, lysates from three different stages of prostate cancer from many patients were immobilized: normal prostate epithelium, prostate intraepithelial neoplasia, and invasive prostate cancer. They were then able to detect and quantify the phosphorylation states of a variety of proteins, including Akt and ERK, throughout the progression of the disease. This technique was sensitive, precise, and linear enough to detect even the smallest changes between samples, and they were able to show a positive correlation between phosphorylation state and prostate cancer progression. This method shows great promise going forward, but like the analytical microarrays, it is limited by the availability and specificity of commercially available antibodies.

Microarray fabrication

Protein purification

As mentioned before, one of the largest hurdles in microarray technology has been the high-throughput purification of protein samples. Unlike DNA, which can be synthesized *in situ* with the same properties as in the cell, proteins cannot be synthesized *in vitro* at high efficiency because of their complex chemistry. Therefore, new technologies had to be developed to accommodate the need for very pure and functional proteins.

Recombinant antibodies have provided a useful alternative to the traditional hybridoma-based technology used for high-quality antibody purification. Phage display was one of the earliest techniques used for recombinant antibody production. Using antibody-fragment encoding genes (VH and VL) and bacteriophage capsid gene fusion, large sets of human antibody libraries are stored in prokaryotic systems, where they can be easily expressed by phage infection (Carmen *et al.*, 2002). After expression, they are displayed in the phage capsid and purified using column chromatography. These antibodies also have the advantage of not containing the Fc domain that intact IgG antibodies contain, therefore eliminating non-specific binding to the Fc receptor (Knappik *et al.*, 2009). More recently, other approaches have been developed as well for the production of recombinant antibodies, including the eukaryotic expression system of yeast display (Chao *et al.*, 2006). Using yeast is beneficial in maintaining some of the post-translational modifications that are absent when using bacteria.

The development of functional microarrays faces a large hurdle as well because of the large quantities of highly purified proteins that are needed. Not only do the proteins need to be free of contaminants, they also require conditions that facilitate proper folding, native post-translational modifications, and require optimal physical conditions during the purification. To this end, large-scale purifications have been developed using both yeast (*Saccharomyces cerevisiae*) and bacteria (*Escherichia coli*) expressions systems (Zhu *et al.*, 2001, 2007; Chen *et al.*, 2008; Jeong *et al.*, 2012). Several advantages exist in producing human proteins in yeast rather than in bacteria: (1) higher solubility; (2) higher yields of large proteins (e.g. >50 kDa); (3) better preserved conformation; (4) less immunogenicity; and (5) they contain many, but not all, of the native post-translational modifications that a prokaryotic system cannot provide. While both can utilize batch purification in a 96-well format to purify thousands of proteins at a time, many labs cannot undergo such a large endeavour.

In 2004, Ramachandran *et al.* (2004) developed a novel technique called the nucleic acid programmable protein array (NAPPA). Plasmid DNA is biotinylated and immobilized on a streptavidin surface adjacent to an affinity-tag antibody (e.g. GST). The GST-tagged protein is then synthesized via *in situ* transcription/translation and immediately binds to the antibody. This allows for the generation of over ten thousand proteins on one slide without the tedious cloning and purification steps, and slides can be stored long term (Ramachandran *et al.*, 2004). However, the array is not 'pure' as the proteins are colocalized with the DNA and can provide some unwanted interactions. Alternatively, the PISA method (protein *in situ* array) bypasses the immobilization of DNA on the array by using free DNA in a reaction mixture on an affinity-tagged surface (He *et al.*, 2001). Angenendt *et al.* (2006) created an array containing over 13,000 spots using a robotic system to accurately mix sub-nanolitre reactions containing the necessary transcription/translation machinery. After the protein is translated in the nanodrops, they bind to the surface of the slide and the DNA is washed away, leaving a 'pure' array. While these cell-free methods of creating protein microarrays eliminate the need for large-scale cloning and purifications, they have not been widely utilized because of the low protein yield, the absence of post-translational modifications, and the difficulties in producing large (>60 kDa) proteins.

Surface chemistry

Once the proteins have been purified, the solid support in which they are immobilized is another factor to be considered. An optimal surface is one that allows for proteins to be tightly bound while still maintaining their three-dimensional conformation, yield high signal-to-noise ratios, and have a long shelf life. Popular types include adsorption, diffusion, covalent immobilization, affinity capture, and metal and are summarized in Table 5.2.

In the early life of this technology, glass slides coated with polyvinylidene fluoride (PVDF), nitrocellulose, or polystyrene were used. Through simple adsorption, proteins were easily arrayed (Stillman *et al.*, 2000). However, because these surfaces are relatively soft, they allowed protein spots to spread out, decreasing the overall density of each spot. It was also found that many applications generate high background for a low signal-to-noise ratio. Another approach uses an agarose/polyacrylamide gel to immobilize proteins in their native conformation and the hydrophilic matrix prevents the lateral movement found in the PVDF-coated membranes. Because of the restriction in movement of the proteins, the size of each spot is decreased, resulting in a greater number of total proteins allowed on the array (Charles *et al.*, 2004). Aldehyde- or epoxy-derivatized glass surfaces are a third

Table 5.2 Summary of microarray surfaces

Characterization	Surface	Properties
Adsorption	Polyvinylidene fluoride (PVDF), nitrocellulose, polystyrene	High background signals in certain assays; low cost
Diffusion	Agarose/polyacrylamide gel, hydrogel	Weak protein immobilization; preservation of protein conformation
Covalent immobilization	Aldehyde, NHS, epoxy, carboxylic ester, etc.	Irreversible protein immobilization; good for covalent reactions; random protein conformation
Affinity capture	Ni^{2+}-NTA, streptavidin, glutathione	Can control protein orientation
Metal	Gold, silver, steel, etc.	Conductive surface; compatible with SPR, OI-RD, and mass spectrometry detection

popular surface that forms covalent bonds with the printed proteins. These provide strong binding interactions and relatively low backgrounds, but protein orientation is randomized (MacBeath *et al.*, 2000). If protein orientation is critical, surfaces coated with streptavidin, glutathione, or nickel allow for biotin-, GST-, or His-tagged proteins, respectively, to bind tightly in a non-covalent manner. By having all the proteins in the same conformation, a large increase in signal-to-noise ratio can be achieved (Zhu *et al.*, 2001). The final surface that is currently being used involves the adsorption of proteins onto a surface coated with gold, silver, or steel. In the case of the gold coated arrays, porous gold is deposited onto a bare gold surface and the substrate is patterned with methyl and carboxy-terminated SAMs (self-assembled monolayers). This creates well-defined hydrophilic spots (carboxy-terminated) where the substrates are covalently attached, and a superhydrophobic background is formed (methyl-terminated) that minimizes protein adsorption from biological solutions. Proteins arrayed in this manner facilitate MALDI-TOF detection of protein arrays (Evans-Nguyen *et al.*, 2008). This type of array is also compatible with surface plasmon resonance spectroscopy (SPR) (Huang *et al.*, 2006).

Microarray printing

Careful placement of purified samples is crucial when fabricating microarrays. Spots must be homogenous and dense, yet spatially discrete, therefore robotic systems are often used to 'print' a surface with samples. There are two methods of spot formation, contact printing and non-contact printing. Each of these methods can also be done in serial (one sample at a time) or in parallel (multiple samples at a time). In this next section we will discuss the wide variations in these methods as well as their respective advantages and disadvantages (Barbulovic-Nad *et al.*, 2006).

Contact printing involves direct contact between a printing device and the substrate. The printing device first comes in contact with a sample, which adheres to the apparatus via surface tension. It then physically contacts the solid surface, where the sample is attached covalently or non-covalently, depending on which surface is used. These technologies use a variety of tools, including solid pins, split pins, microstamps, and nano-tips. Initial microarrays used a single pin for fabrication, but that was quickly evolved into an array of pins to increase throughput. Pin printing is considered serial printing, while microstamps deposit large amounts of sample in a parallel fashion. Nano-tip printing is based on atomic force microscopy (AFM) and can generate arrays with submicron spots.

Pin printing is one of the most common forms of microarray fabrication. There are a variety of pins that can be used, each with its own characteristics. Solid pins are the most simple and were the first to be used. They consist of a solid piece of metal that is dipped into the sample solution to 'load' the tip. Capillary force action allows for the sample to be loaded as a small bead on the tip of the pin, and it is deposited on a solid surface once the pin comes in contact with it. The pins vary in shape, from convex to flat to concave. One loaded sample can print only a few spots, which is why these pins are generally only used for low-density arrays. They do offer a few advantages, however. If sample solutions are very viscous, they can clog the split-pin types, which is not an issue for solid pins. Cleaning of these pins is also much easier, so arrays are typically very reliable between batches.

The most common type of pin that is used today is the split pin. These pins contain small (60–100 μm) slits in the centre that allow a greater volume of sample to be loaded in each dip with capillary action, greatly increasing throughput. However, because the same pins are often used to handle multiple samples, they have to be carefully washed between samples. This typically includes a sonication bath step between samples to assure that all sample liquid is removed. Some disadvantages need to be considered, however. Although made of metal (typically solid steel, tungsten, or titanium), the very fine tips can still be subject to deformation from the tapping force and lead to uneven spotting. Dust and contaminants as well as highly viscous solutions can lead to clogging as well. This can be resolved by using a wider slit, but spot size reproducibility is affected. Choosing the right pinhead will depend on the sample, so careful consideration must be applied to achieve optimal conditions.

The actual printing system involves a pin head (holding anything from 16 to 96 pins) attached to a translation system, a sample plate, and a flat plane with the carefully positioned solid surface substrates (Fig. 5.2). The pins are free to move vertically while in the pin head, so there is flexibility when they come in contact with the solid surface. The translation machinery moves the pin head to the source plate, where the pins are lowered to collect the sample through capillary action. The sample is held by surface tension either on the tip (solid pins) or within the capillary (split pins). The pin head is then positioned over the solid surface and gently, but reproducibly, tapped, dispensing the sample onto the solid surface where it is immobilized. Spot sizes are defined by many factors, including sample viscosity, pin velocity, substrate surface properties, precision of the robotic controls, humidity, and temperature. Although pin printing represents the most popular form of microarray fabrication, it can be time-consuming. For example, a single 384-well source plate printing can take up to 6.4 hours, so other spotting methods were developed to overcome this limitation.

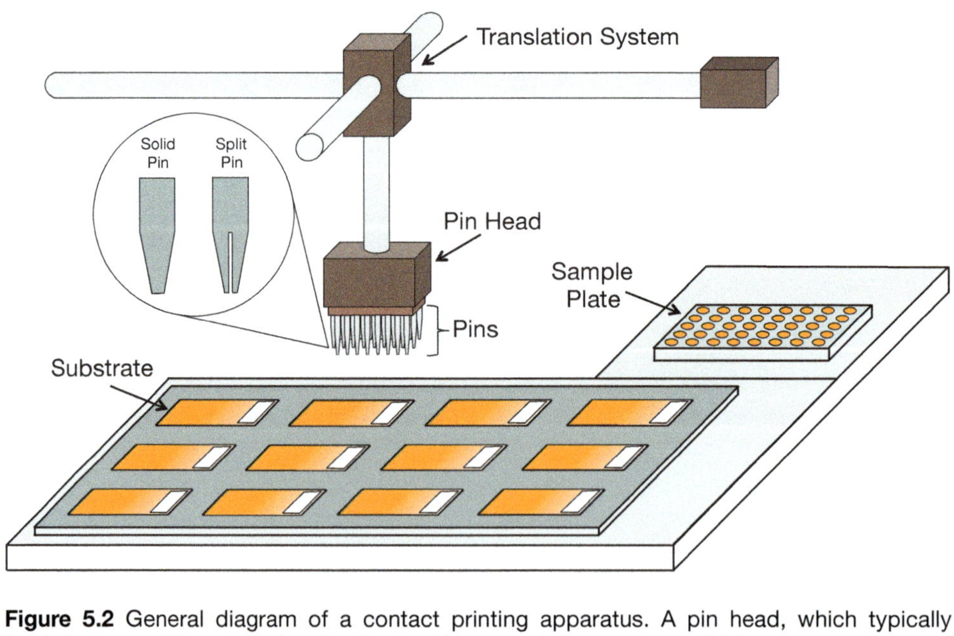

Figure 5.2 General diagram of a contact printing apparatus. A pin head, which typically holds between 16 and 96 pins, is attached to a translation system, which can accurately and reproducibly move in 3-D space. The pins are first dipped into a sample plate, where the sample adheres to the pins with capillary action. The pins are then moved above the arranged substrates and spotted by physically tapping the surface. As a serial printing technique, this is generally time-consuming, as a single 384-well source plate can take over 6 hours to fabricate.

Microstamping is a serial contact printing technique that is both simple and cheap enough to be performed in any laboratory, while allowing for the high-throughput fabrication of protein microarrays. The basic protocol involves a protein sample being adsorbed onto a patterned surface and then transferred to a substrate using physical contact. Obtaining good contact is critical in the sample spotting, so elastomeric materials, such as poly(dimethylsiloxane) (PDMS) are generally used because they conform to solid surfaces under an applied load. These elastomeric stamps are produced using a micromoulding technique that requires a single photolithography step. A master mould is created by photolithography, followed by casting of an uncured elastomer onto the mould. After curing, the stamp is removed from the mould, where it can be used to adsorb to samples. Through this process, many disposable stamps can be easily fabricated, eliminating the need for the washing between samples that is required for pin printing. The major disadvantages of the stamping method involve the amount of sample that can be transferred. The sample volume transferred from stamp to substrate is not well controlled and depends greatly on the surface and sample properties. Moreover, initial sample volumes are much larger than in pin printing because only a small fraction adsorbs to the stamp and is deposited on the substrate, which can be a serious limitation when dealing with precious samples (such as patient tissues). Spots are also much larger than what can be produced using pins (300 μm vs. 60 μm), which leads to decreased spot density. Likewise, the washing of microstamps for reuse is more difficult and tedious than for pins due to the non-specific adsorption that can occur.

Inking is the first step performed in microstamping and begins by treating the hydrophobic PDMS stamp with a hydrophilic film to insure sample adherence. Similar to a standard ink and rubber stamp, the microstamp is then simply inked by dipping into a sample well, where hydrophilic interactions mediate adsorption onto the stamp. In order to obtain uniformity in the amount of sample that is adsorbed to the stamps sites, some new techniques have been developed. Injecting devices called piezoneedles have been used to deliver equal amounts of DNA solution to stamp sites. This technique exhibits good control of spot size and is able to produce multiple arrays from a single inking step. Another technique involves loading the sample solution into wells on the back of a stamp. The solutions flow through microchannels down to the stamp tips through surface tension, and are deposited onto the substrate when careful pressure is applied.

Two types of stamping are used, direct and indirect. Direct stamping involves the stamp directly interacting with a biological sample before being brought into physical contact with the substrate, typically for a few seconds to ensure good conformation. The stamp is then removed, leaving the sample behind. Using the necessary speed, contact angle, surface roughness and spacing between stamp sites is needed to achieve uniform shape and volume in spots. One type of direct stamping technique involves a microfluidic design, which contains both inlet and outlet ports for which the biological sample can be channelled. A stamp is pressed onto the substrate and the sample solution flows through the channel and prints a spot onto the substrate. This process can be used to print incredibly concentrated samples because the number of desired molecules can be immobilized to the substrate without washing away unwanted materials.

In indirect stamping, the stamp is inked with chemical groups or self-assembled monolayers (SAMs) rather than with biological samples. Typically SAMs are patterned on a gold substrate with alkanethiols, in which a head group adsorbs onto the gold surface, followed by a long polymer carbon chain, and ending with a reactive tail group. The stamp first prints an array onto a gold substrate using a particular alkanethiol (AT1) using traditional contact methods. Next, the remaining un-modified gold surface is coated with a secondary alkanethiol (AT2). Samples are designed so that they selectively react only with the AT1 head group and an array is formed. This type of printing is not suitable if more than one sample is to be arrayed, as non-specific adsorption and cross-reactivity can easily occur.

Nano-tip printing is the third method of contact printing techniques that is currently employed for microarray production. These technologies are able to print spots at the sub-micron level through the use of atomic force microscopy (AFM). AFM nano-tips are used in two ways: (1) adding a sample or sample binding molecules to a substrate (dip-pen lithography); or (2) removing SAM molecules from a coated surface (ATM grafting).

Dip-pen lithography uses a similar schematic to indirect sampling, but can do so at a much smaller, sub-micron level (compared to spots 300 μm in diameter obtained through microstamp printing). A sample that binds proteins is transferred to the substrate with the nano-tip in a patterned array. The remaining areas are blocked with molecules resistant to protein adsorption, as before. The arrays are finally treated with a biological sample, which only binds to the compatible surface. This has been shown to create monolayers only 30 nm wide.

AFM grafting, conversely, uses the nano-tip to remove one SAM layer (consisting of molecules resistant to protein adsorption) at selected sites, followed by treatment with reactive SAM molecules to fill in the gaps. This is useful when nano-patterns need to be modified *in situ*, eliminating the need to repeat the entire process. The utility of this fabrication method is more limited than the other two for a few reasons. One, it is a serial technique, so it is much slower than microstamping. The slow speed reduces fabrication efficiency as well, as sample volumes must be very small because they tend to dry up during the printing process. Also, as the individual spots approach protein size levels, non-specific binding can become a large issue. Finally, this approach, like indirect stamping, can generally only accommodate very few biological samples at a time, compared to the thousands of unique proteins that can be printed using either direct stamping or pin printing techniques. However, the greatest advantage arises from the fact that both the printing and detection of the microarrays can be performed on the same instrument, as AFM has been shown to be able to detect protein-biological sample interactions (as described later in label independent detection methods).

In contrast to contact printing, which all involved direct physical interaction between printer and substrate, non-contact printing approaches obviously do not. These techniques range from photochemistry-based methods to fluid droplet dispensing methods, but all share the same advantages: reduced contamination and higher throughput. By separating the printer from the substrate at all times, cross-contamination between samples is dramatically reduced. Additionally, the need to clean the printing devices between samples is also eliminated. Moreover, non-contact printing techniques offer the potential to increase throughput over other techniques. Many can also deposit solutions in parallel, allowing entire arrays to be produced simultaneously.

Photochemistry microarray printing involves chemical treatment of the substrate, followed by UV light exposure through photomasks. Two types of photochemistry techniques are used, photolithography and direct photochemical patterning. In the photolithography approach, a photoresist layer is coated onto a substrate surface, followed by UV light exposure through a patterned photomask (Fig. 5.3A). This forms micrometre-sized open

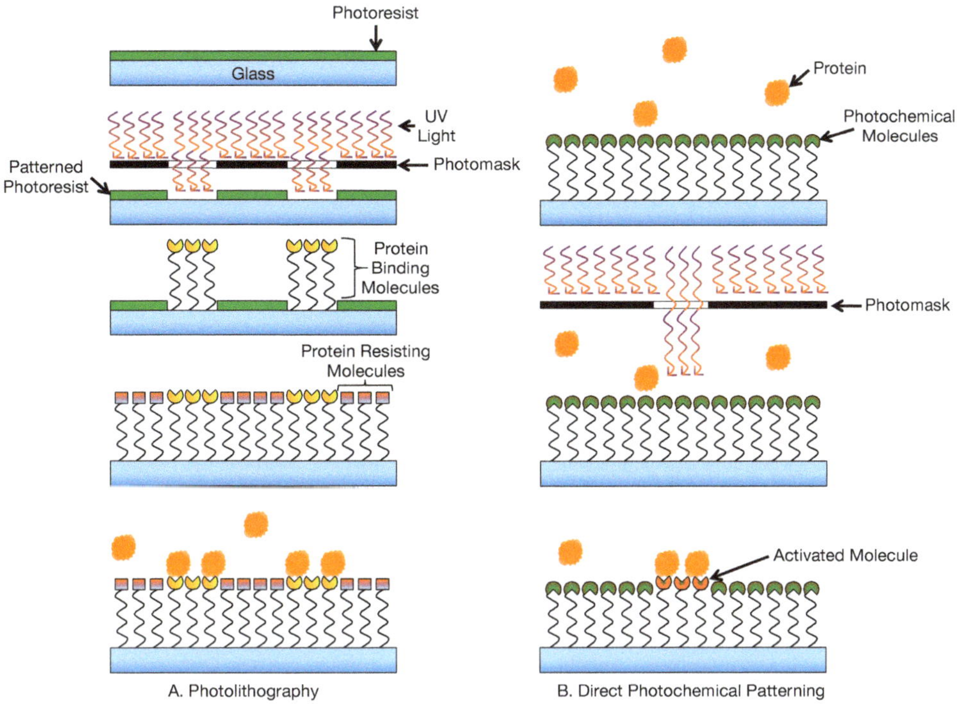

Figure 5.3 Photochemical microarray printing as a non-contact printing approach. (A) Photolithography printing first involves coating a glass surface with a photoresist coating. Illuminating the surface with UV light through a patterned photomask creates a patterned photoresist. The ablated spots are then treated with protein-binding molecules. After the remaining photoresist is removed, the remaining glass surface is coated in protein resisting molecules. Finally the protein mixture is added, where it can adhere to the array in a patterned format. (B) A similar approach, direct photochemical patterning, uses glass slides coated with photochemical molecules first in the presence of biological samples. UV light is again shone through a patterned photoresist, specifically activating these molecules, and allowing the proteins to bind. Both of these methods are limited in the diversity of samples that can be spotted, but benefit from the speed in which they can be fabricated.

regions where adhesion-promoting molecules can then bind. The remaining photoresist layer is removed and the substrate is incubated with adhesion-resisting molecules. Direct photochemical patterning is similar to photolithographic techniques, but it does not require a photoresist layer (Fig. 5.3B). Substrates are coated with photochemical molecules and activated by UV light through photomasks, where they can subsequently bind to biological molecules. These techniques are high throughput, but again, are limited in the variety of samples that can be patterned at once.

The other type of non-contact printing technique involves dispensing droplets of protein sample onto a substrate. The three most commonly used approaches are inkjet printing, electrospray deposition (ESD), and laser writing.

In an effort to reduce the cost of microarray printing machinery, attempts have been made to find cheaper alternatives in commonly available products. Oftentimes, commercially available printers are modified to dispense biological materials instead of ink. There are two commonly used inkjet printing techniques, thermal and piezoelectric. Thermal printers utilize resistive heaters (with temperatures reaching upwards of 200°C) for sample droplet dispensing, while piezoelectric printers use piezoelectric actuators to dispense droplets onto the substrate surface. Although an attractive technique, there are several disadvantages. For starters, commercially available printers are not designed to print on glass slides, so they are limited to flexible membranes such as cellulose, nylon, and nitrocellulose. These surfaces can lead to smearing, and therefore contamination between samples. Secondly, the nozzles used in inkjet printing tend to produce 'satellite' spots surrounding the spot of interest, reducing printer resolution. Cleaning of the nozzles is also difficult, especially in piezoelectric printers. Finally, samples can experience high shear rates and high temperatures, regardless of the method used, which can result in the denaturing of biological samples. Some studies have shown an ability to overcome some of these problems, namely the cleansing issue, by using deposition heads with large numbers of top loaded reservoirs (Gutmann *et al.*, 2004). Many droplets from different samples can therefore be printed simultaneously.

Similar to inkjet printing, electrospray deposition (ESD) is a technique that has been modified from an existing application to be used with biological samples. ESD has been typically used to deposit thin films of polymers, semiconductive ceramics, and radioactive sources, but many have begun using it to deposit biological samples on a substrate for microarray fabrication. The general technique uses a dielectric mask placed between the capillaries containing biological samples and the substrate. An electrostatic field in produced between the capillary and the substrate, driving the solution out of the capillary nozzle. As the inverse charge is applied below the surface of the substrate, the biological samples are attracted to the surface, but can only pass through the holes of the dielectric mask, where they are deposited. Each capillary is filled with a different substrate, and sequentially activated after movements of the mask to different spots on the array. This allows for fast and parallel fabrication of microarrays, as well as the production of very small spots (2–6 μm). However, although spot size can be small, the distance between spots (called 'pitch') is relatively large, on the order of 1 mm. This limits the overall number of spots that can be arrayed onto a single substrate slide. Another limitation arises from differences in droplet distribution from the spraying technique. Generally, only those spots directly below the capillary produce high-density spots, but surrounding spots are generally quite irregular in regards to density. However, because the pitch is so large, cross-contamination is generally not an issue. High shear rates can be damaging as well, as the solution becomes charged due to the electrostatic field. This field can cause deformations in proteins, as well as alterations in solution pH, so it is important to take into consideration when selecting samples to use with this method.

The final method of non-contact printing uses laser ablation to produce protein microarrays, either directly or indirectly. In direct writing, a sample is mixed with glycerol and buffer and coated on a quartz disc. As a pulsed laser is scanned across the disc and local evaporation occurs, producing microscopic droplets that fall to the substrate surface. The droplets produced are smaller than one can get using traditional pin methods (50 μm) and does so with very little sample. Using this technique, over 16,000 arrays can be produced with only 500 nl of starting material. Indirect writing is a process that is similar to many of the other 'lift-off' techniques mentioned before. A photoresist layer is placed on top of a substrate and is selectively removed with a photomask and laser, leaving patterned pockets with which to apply biological samples.

While all these non-contact printing methods hold great promise for the production of sub-micron spots, all require high shear, high temperatures, or both. Therefore careful consideration must be had when choosing to use these approaches.

In summary, there have been many attempts by various labs to create a fabrication system that optimizes uniformity, minimizes sample volume, and does so without contamination of the biological samples. Pin printing remains the most popular player in this process, as it overall provides the most reproducible arrays with little maintenance, albeit with the highest cost. Many of the ablation techniques can produce very small spots, but are severely limited in the number of unique samples that can be spotted on one array. Non-contact techniques eliminate many of the contamination concerns present in pin and stamp printing, but at the cost of high temperatures and shear forces. As these technologies inevitably advance, these shortcomings have promise to be overcome.

Detection methods

The detection methods used for protein microarrays are another important design parameter. There are two classes of detection: label dependent and label free.

Label dependent

Several types of label dependent detection methods have been developed and optimized (Fig. 5.4A). The first is the use of fluorescent dyes with narrow excitation and emission spectra, such as Cy3 or Cy5. These are most commonly used for their convenience and wide detection range. Furthermore, they can also provide a multicolour system for multiplex assay design (Fu *et al.*, 2007). Enzymatic methods can also be used to enhance signal amplification. The most common is horseradish peroxidase, but others include a modified rolling circle amplification (RCA) label that uses DNA primers attached to antibodies to create binding partners for fluorescent, complementary oligonucleotide probes. This is especially useful for the detection of low abundance proteins, where the sensitivity of chemiluminescence and traditional fluorescence is too low (Schweitzer *et al.*, 2002). Other assays that involve enzymatic reactions on the immobilized proteins can use radioisotopes (e.g. ^{32}P, ^{33}P and ^{14}C) for *de novo* detection. While this can provide superior signal-to-noise ratios, and can be the only reliable detection method for modifications without high-affinity antibodies, many oppose this method due to safety concerns (Chen *et al.*, 2006).

Label free detection

Although useful, labelling processes can alter protein activity, so many label-free methods have been developed (Fig. 5.4B). Mass spectrometry has been used for the detection of ligand binding, using MALDI-MS, SELDI-TOF-MS, and MALDI-TOF-MS approaches (Gavin *et al.*, 2005; Diamond *et al.*, 2003; Evans-Nguyen *et al.*, 2008). This approach can be done quickly and simply and with very little sample, and can also be used to directly detect analytes bound from complex samples such as urine, serum, plasma, and cell lysates (Zhang *et al.*, 2012). Atomic force microscopy (AFM) is another approach that can identify analytes bound to an array by detecting a change in height of the samples on the array, which leads to measurable binding interactions (Yan *et al.*, 2003). Another advantage of label-free methods arises from the ability to detect real-time dynamics of protein interactions. Surface plasmon resonance (SPR) has been used to analyse bimolecular interactions in real-time and has been adapted for the protein microarray (Wegner *et al.*, 2003; Unfricht *et al.*, 2005).

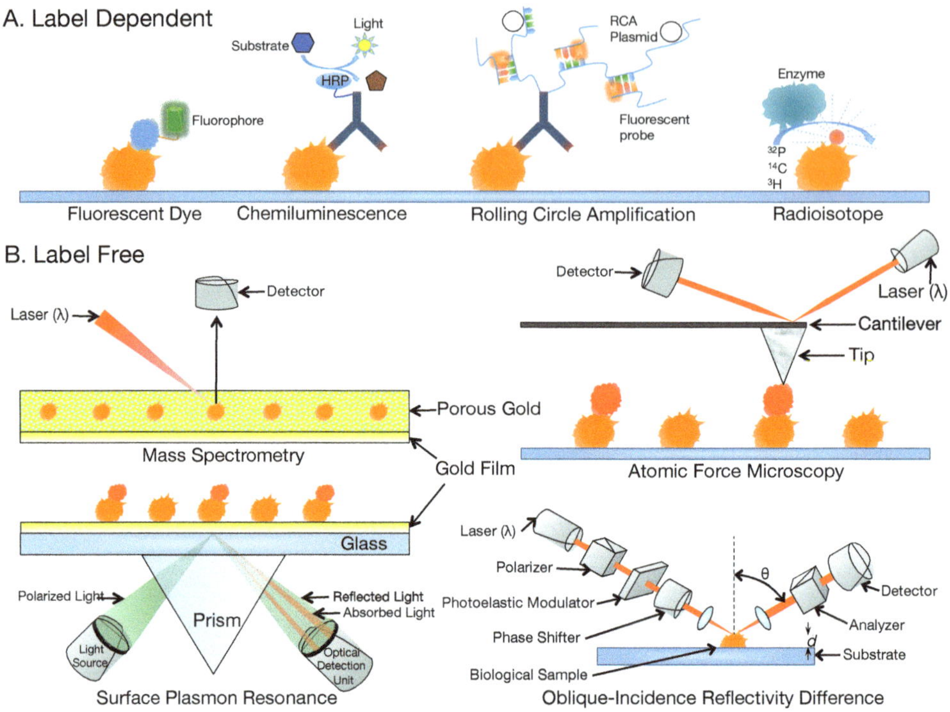

Figure 5.4 Detection methods used in protein microarray experiments. (A) Label-dependent detection can alter protein activity, but provide great sensitivity. Fluorescent dyes are among the most commonly used, and can utilize multiple colours for detection of multiple interactions (i.e. acetylation and protein levels, concurrently). Chemiluminescent techniques can also be used, such as horseradish peroxidase (HRP). RCA approaches are useful with low abundance proteins, as they can greatly amplify the signal. Radioisotopes are useful for *de novo* assays and are preferred when specific antibodies are not available. (B) Label free methods eliminate the possibility of altering protein activity, but at the cost of expensive equipment and lower throughput. Mass spectrometry approaches can be used with complex mixtures. Atomic force microscopy (AFM) can detect ligand binding by sensing a change in height as a cantilever moves across the array. Surface plasmon resonance (SPR) is used to detect binding in real-time by observing changes in incidence light from the resonance of the surface plasmon when a ligand binds to and dissociates from its target. Oblique-Incidence Reflectivity Difference (O-IRD) detects changes in reflectivity between p- and s-polarizations, and can likewise be used to detect changes in biological samples from ligand binding in real-time.

Incident light can resonate with plasma on a metal surface in total internal reflection, causing resonance signals to change when ligands bind to (and dissociate from) proteins on the array surface. In this way, binding events can be monitored in real-time and kinetic parameters can be calculated. As an extension of traditional SPR techniques, SPR imaging (SPRi) combines the sensitivity of SPR with the spatial capabilities of imaging. A CCD camera is used to capture the light reflected from the entire surface and can be quantified by subtracting the background signal from the image taken before the experiment begins. As samples flow over the surface of the array, multiple images are acquired, and reflectivity changes can be quantified in real-time with a visual output as well (Kodoyianni, 2011). Similarly, the oblique-incidence reflectivity difference (O-IRD) approach allows for the very sensitive detection in the changes in reflectivity between the p- and s-polarizations. It is also used for acquiring kinetic parameters by detecting the tiny changes in the physical properties of a biological sample, such as thickness and density (Landry *et al.*, 2008).

Basic research applications

Among the growing fields of scientific research that utilizes the protein microarray, basic research has been at the forefront of this technology. There is an incredible amount of diversity in the applications of this technology, which we will highlight in this section, and while these studies all involve the same general schematic, they all arrive at vastly different conclusions (Fig. 5.5). These studies include the detection of protein–binding interactions (protein–protein, protein–lipid, protein–peptide, protein–DNA, protein–RNA, protein–small molecule, and protein–glycan), detection of post-translational modifications (phosphorylation, ubiquitylation, acetylation, nitrosylation, sumoylation, and glycosylation), and profiling of monoclonal antibody specificity.

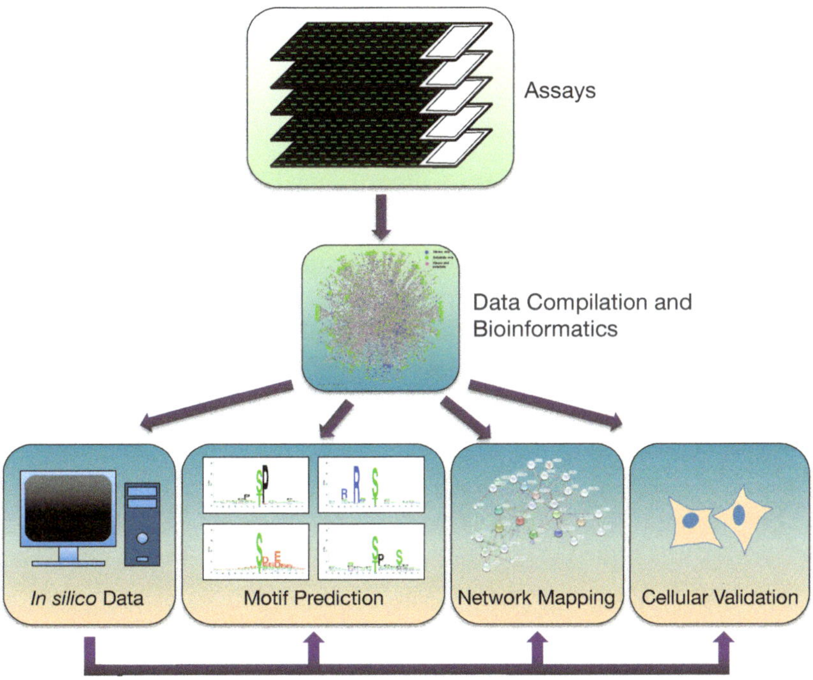

Figure 5.5 General schematic for microarray studies. A typical microarray study involves three steps: (1) Completion and detection of assays; (2) Compiling data for bioinformaticians, where biologically relevant connections are first made; (3) validation and further observations. *In silico* conclusions can be drawn through statistical algorithms, but all require further validation. Many motifs types are commonly predicted from the commonalities between hits, such as binding motifs and PTM motifs. Interactions can be combined with known data to formulate dense networks. The final step generally requires some sort of *in cellulo* or *in vivo* validations. One common weakness of all microarray studies arises from false-positives and false-negatives that can arise from any *in vivo* system, so a combination of bioinformatics and cellular validation studies helps eliminate these biases.

Detection of protein-binding interactions

One of the most basic fundamentals of cell signalling involves the physical interaction between biological molecules. Identifying these interfaces is critical in developing our understanding of how our increasingly complex world works. Many early studies have used two-hybrid screens to elucidate protein–protein and protein–DNA interactions (Vidal *et al.*, 1996; Young, 1998; Joung *et al.*, 2000). This technique is a popular approach due to its low-tech nature and scalability, therefore is often used as a first step in identifying novel interactions. However, there is criticism due to the many false positives and false negatives that can arise for a variety of issues. As fusion proteins, some interactions may be inhibited due to the non-native nature of the protein. Hybridization also takes place in the nucleus, limiting the breadth of the studies to those proteins that localize to the nucleus. Protein complex purification coupled with mass spectrometry analysis is another approach that has been established to identify novel protein–protein interactions (Krogan *et al.*, 2006). While this method has proved to have high fidelity, it is both time-consuming and not suitable for low abundance proteins. These issues, among others, contributed to the development of protein microarray technology for the study of protein–binding interactions. These arrays can be screened in parallel quickly, reproducibly, and in high throughput. One drawback lies in the inherent *in vitro* nature of the assay itself, as interactions identified on the array may not occur in the context of a cell. Therefore, hits must be further characterized in order to be truly validated. Although not without some disadvantages, protein microarrays have become one of the most widely used technologies for the discovery of protein–binding interactions (Table 5.3).

Protein–protein interactions

MacBeath and Schreiber (2000) performed the seminal interaction study using this technology in 2000. By utilizing the same robotic equipment used to print DNA microarrays, they were able to create high-density protein microarrays and identify protein–protein interactions, kinase substrates, and small molecule targets. As a proof of principle, they first

Table 5.3 Applications of microarrays to detect protein-binding interactions

Interaction type	Array content	Type of probe	Reference
Protein–protein	60 EBV viral proteins	Human protein	Zhu *et al.* (2006)
	4191 human proteins	Viral protein	Shamay *et al.* (2012)
Protein–lipid	Yeast proteome	PI(3,5)P2_2 liposomes	Lu *et al.* (2012)
Protein-peptide	159 human SH2 and PTB domains	Peptides	Jones *et al.* (2006)
Protein–DNA	282 yeast TFs	DNA motif	Ho *et al.* (2006)
	4191 human proteins	DNA motif	Hu *et al.* (2009)
Protein–RNA	9400 human proteins	Coding and noncoding RNAs	Siprashvili *et al.* (2012)
	Yeast proteome	BMV RNA loop	Zhu *et al.* (2007)
Protein–small molecule	Yeast proteome	Small molecule inhibitors/ enhancers of rapamycin	Huang *et al.* (2004)
Protein–glycan	Yeast proteome	Lectins	Kung *et al.* (2009)
	94 lectins	Live mammalian cells	Tao *et al.* (2008)

sought to prove that this technique allowed for the functional properties of the immobilized proteins to remain intact. Three pairs of known interactors were used to test this hypothesis: Protein G and immunoglobulin G (IgG); p50 (of the NF-κB complex) and its inhibitor I-κBα; and the FKBP12-rapamycin binding (FRB) domain of FKBP-rapamycin-associated protein (FRAP) and the human immunophilin FKBP12 (12 kDa FK506-binding protein). The first protein of each pair was immobilized on a glass surface followed by incubation with a fluorescently labelled pair. Each pair utilized a different fluorophore and they were able to show that they each bound only to their respective interactors, even simultaneously. Another test involved printing 10,800 spots onto a glass slide, with 10,799 of them containing Protein G and one specifically printed with FRB. Using a mixture of both blue (IgG) and red (FKBP12) probes, they were able to show only one spot fluoresced red, only where the FRB was spotted. This initial study signified that the immobilization of the proteins onto derivatized slides did not affect their binding properties.

Shortly thereafter, Zhu *et al.* (2001) analysed protein–protein interactions using test ligands labelled directly or indirectly with fluorescent dyes. Using the first 'proteome' microarray mentioned before containing approximately 5800 recombinant yeast proteins (>85% of the yeast proteome), they were able to identify the binding partners of calmodulin (CaM). After first incubating the microarray with biotinylated bovine CaM, the proteins were detected with Cy3-labelled streptavidin. Apart from identifying six known CaM targets, 33 new CaM binding partners were discovered. Further bioinformatic analysis revealed a shared consensus motif between 14 of the 39 binding partners: (I/L)QXXK(K/X)GB, where X is any residue and B is a basic residue. Not only can targets be identified easily through this approach, motifs can also be deduced, further expanding the usefulness of these arrays.

In a related study, Popescu *et al.* (2007) developed a protein array consisting of 1133 *Arabidopsis thaliana* proteins to globally identify binding partners of CaM or calmodulin-like proteins (CML). Instead of using biotinylated CaM, they used CaM amino-conjugated to Alex Fluor 647 to identify 173 different targets that bound to the three CaMs and four CMLs. Like the yeast study, a large number of previously known CaM targets were identified, as well as a diverse group of transcription factors (TFs), receptor and intracellular protein kinases, RNA-binding proteins, F-box proteins, and proteins of unknown function.

Protein–lipid interactions

Apart from identifying CaM binding partners, Zhu and colleagues also identified the binding partners for six liposomes containing different phosphatidylinositides (PIs) on the yeast proteome array (Zhu *et al.*, 2001). The liposomes contained 1% N-(biotinoyl)-1,2-dihexadecanoyl-*sn*-glycero-3-phospho-ethanolamine, triethylammonium salt (biotin-DHPE) and were similarly detected using a streptavidin-labelled Cy3 fluorescent dye. The six liposomes were able to identify 150 different protein targets, including integral membrane proteins, anchored proteins, prenylated proteins, and lipid metabolism enzymes. The binding interactions could also be characterized as 'strong-binding' or 'weak-binding' by comparing the fluorescent signal obtained by the liposomes to the fluorescent signal obtained from GST probing (this standard technique uses a Cy5 labelled antibody targeting the tag of the immobilized proteins on the array, yielding the relative protein concentration at each spot). Through this analysis, they were able to correlate strong binders with membrane-associated proteins and that strong phosphoinositide-binding proteins preferentially bound a specific PI. Not only could binding interactions be identified, but conclusions can be drawn concerning the binding affinities of the various PIs. This technique of comparing the binding/modification levels to the protein levels is one that is utilized often, and is another reason why protein microarrays are one of the more versatile platforms available.

More recently, Lu *et al.* (2012) developed another fluorescent liposomal assay to identify binding partners for phosphatidylinositol 3,5-bisphosphate ($PI(3,5)P2_2$) using the same yeast proteome array used by Zhu *et al.* (2001). A special non-quenched fluorescent (NQF) liposome was developed that eliminates the need to lyse the liposomes in order to see fluorescent signal. Using these modified liposomes containing ($PI(3,5)P2_2$), 162 binding proteins were identified, 22 of which were novel and shared similar functional roles with known interacting partners, including vesicle-mediated transport machinery, GTPases, and protein kinases. Collaboration with a bioinformatician revealed a novel motif, HRDIKP(E/S)N(I/L)LL, that was significant among the 162 novel ($PI(3,5)P2_2$) binding partners. While the hypothesis was that this was not the site of PI binding, it represented a novel motif among PI-interacting proteins. This approach exemplifies a beneficial improvement over traditional fluorescent liposomal nanovesicles, as it can be used on microarrays, and has promising applications in future high-throughput studies of protein–lipid interactions. It is also important to note that with large studies such as this, it is crucial to have strong relationships with bioinformaticians to aid in the analysis of the large numbers of data that are produced.

Protein–peptide interactions

Jones and colleagues developed a unique array that used protein domains instead of full-length proteins in order to investigate protein–peptide interactions deemed important in the ErbB family of receptors (Jones *et al.*, 2006). The array was printed with 106 Src homology 2 (SH2) and 41 phosphotyrosine binding (PTB) domains, along with several tandem domains. Arrays were incubated with 61 peptides representing tyrosine phosphorylation sites on the four ErbB receptors. In order to quantitatively measure the protein–peptide interactions, eight concentrations of each peptide (between 10 nM and 5 mM) were tested, leading to the following insights: (1) the study of these interactions in a non-competitive format revealed high-affinity binding sites for both SH2 and PTB domain, but do not imitate the consensus recognition sequences; (2) ErbB2 recruitment sites are more promiscuous than any of the other receptors; (3) when comparing the highest–affinity interactions, the proteins that bind to EGFR constitute a small subset of those that bind to ErbB3; and (4) at high concentrations of EGFR and ErbB2, the binding becomes much more promiscuous, which may contribute to the high oncogenic potential that has been seen in various cancers.

Protein–DNA interactions

Another common application of protein microarrays is in the characterization of protein–DNA interactions (PDIs). One of the earliest studies screened novel DNA-binding proteins by probing the yeast proteome microarray with fluorescently labelled yeast genomic DNA (Hall *et al.*, 2004). Greater than 200 proteins were identified, and half of those were not previously known to bind DNA. In order to determine whether the novel DNA-binding proteins were not artefacts (i.e. they non-specifically bound *in vitro*, but not *in vivo*), the group tested eight novel proteins by using ChIP/chip. The proteins of interest were tagged with 13 c-Myc epitopes at the C-terminal before treatment with formaldehyde to crosslink the protein to the potentially bound chromosomal DNA. The Myc-tagged protein was then immunoprecipitated from the cell lysates, where the DNA was labelled with Cy5. Five of the eight proteins showed no enrichment in any particular DNA loci, but three proteins (Mtw1, Dig2, and Arg5,6) were found to be specifically associated with DNA loci *in vivo*. The most important of which was Arg5,6, which is a single yeast gene that encodes two enzymes involved in arginine biosynthesis. And while Mtw1 and Dig2 bound to multiple loci, Arg5,6 only bound to one specific loci found in both nuclear and mitochondrial chromatin. Deletion of Arg5,6 altered the transcript levels of both nuclear and mitochondrial target genes, signifying that metabolic enzymes can directly regulate eukaryotic gene expression.

Though shared genomic DNA can be used as probes, individually synthesized short DNA motifs can also be used as probes to interrogate the entire yeast TF repertoire immobilized on a glass slide. Indeed, the Snyder and Johnston groups were the first to establish this approach as reported in a PNAS paper in 2006 (Ho *et al.*, 2006). Using a microarray containing 282 known yeast TFs, they were able to identify interactions with 75 evolutionarily conserved DNA motifs. Again, over 200 PDIs were identified, with greater than 60% of them previously unknown. It was shown that 15 of the proteins bound almost all of the DNA probes non-specifically, while 62 proteins bound to at least one probe. Further analysis identified Yjl103p as a novel DNA-binding protein and the DNA binding motif was found to be $CGGN_{8/9}CGG$. In order to obtain the target genes for this transcription factor, Yjl103p was either overexpressed or deleted before gene expression profiling. Over 500 genes were differentially expressed with overexpression, many of which were involved with carbohydrate metabolism, stress response, or oxidative phosphorylation. This was one of the first transcriptome-wide studies performed on an entire organism and paved the way for the analysis of the human transcriptome.

Following up on the yeast transcription factor study from Ho *et al.* (2006), Hu and colleagues undertook a large-scale analysis of human PDIs using a protein microarray composed of 4191 unique human proteins in full-length (henceforth known as the TF array), including approximately 90% of the annotated TFs and a wide range of other protein categories, such as RNA-binding proteins, nucleotide-binding proteins, transcription coregulators, mitochondrial proteins, and protein kinases (Hu *et al.*, 2009). The protein microarrays were probed with 400 predicted and 60 known DNA motifs and a total of 17,718 PDIs were identified. Many known PDIs and a large number of new PDIs for both well characterized and predicted TFs were recovered, and new consensus sites for over 200 TFs were determined, which doubled the number of previously reported consensus sites for human TFs. Unexpectedly, over 300 proteins that were previously unknown to specifically interact with DNA showed sequence-specific PDIs, suggesting that many human proteins may bind specific DNA sequences as a moonlighting function. To further investigate whether the DNA-binding activities of these unconventional DNA-binding proteins (uDBPs) were physiologically relevant, in-depth analysis was carried out on a well-studied protein kinase, Erk2, to determine the potential mechanism behind its DNA-binding activity. Many *in vitro* and *in vivo* assays were performed, including electrophoretic mobility shift assays (EMSAs), luciferase assays, mutagenesis studies, and chromatin immunoprecipitation (ChIP). These studies showed that the DNA-binding activity of Erk2 was independent of its protein kinase activity and it acts as a transcription repressor of transcripts induced by interferon gamma signalling. Other than Erk2, many other uDBPs showed sequence-specific DNA-binding activity, and many of their consensus sequences are highly similar to those recognized by annotated TFs. This suggests that moonlighting functions of uDBPs based on their sequence-specific DNA-binding activity may be a widespread phenomenon in humans.

Using the same approach, new proteins that may carry enzymatic activity in DNA repair can be discovered. A bacterial proteome microarray composed of 4256 proteins encoded by the *E. coli* K12 strain (≈99% coverage of the proteome) was developed by Chen and colleagues (Chen *et al.*, 2008). In order to identify proteins involved in DNA damage recognition, end-labelled, double-stranded (ds) DNA probes containing abasic or mismatched base pairs were used. Two proteins of unknown function, YbaZ and YbcN, were identified as binding with high affinity to abasic sites and mismatched sites, respectively. Both of these proteins were shown to be base-flipping through the use of a 2-aminopyrine (Ap)-modified DNA approach. The fluorescence of 2-Ap is quenched when properly paired within the duplex, but the fluorescent intensity increases dramatically when flipped into an extrahelical conformation. An increase of more than 1000-fold was observed for both proteins. By probing the same proteome array with purified YbaZ, it was also found that YbaZ tightly interacts with the type IV helicase HelD.

Protein–RNA interaction

It was long thought that RNA was only used for translation (mRNA, rRNA, and tRNA), but with the discovery of over 20 types of RNA, many have been characterized and shown to have a wide variety of functions in biology. Long non-coding RNAs (lncRNAs) are most commonly thought as precursors to short RNA fragments, such as microRNAs. However, recent studies have shown that some of these lncRNAs can be several kilobases in length, and can be conserved extensively at the nucleotide level (Guttman *et al.*, 2009). Like microRNA, some lncRNAs block gene expression by antisense base pairing, but many act *in cis* by regulating mRNA transcription. This is achieved by modifying heterochromatin formation near genomic loci. Nonetheless, there have been relatively few lncRNAs functionally characterized, especially those which appear to act *in trans* to regulate gene transcription. While there are existing methods available to study protein–RNA interactions, they are laborious and require large amounts of cell material. Protein microarrays provide a unique and high-throughput platform from which these interactions can be quickly identified.

Rapicavoli and colleagues were among the first to do so when they sought to further characterize the long non-coding RNA *Six3OS*, which is coexpressed with the homeodomain factor Six3 (Rapicavoli *et al.*, 2011). Six3 plays a pivotal role in mammalian eye development, where it regulates both early eye formation and cell specification to the postnatal retina. Utilizing a human protein microarray, they were able to demonstrate that both mouse and human *Six3OS* was able to bind to five proteins, including Eya1, a homologue of the *eyes absent* gene of *Drosophila*, and Ezh2, a chromatin remodelling enzyme. It was proposed that *Six3OS* modulates the expression of Six3 target genes by acting as a transcriptional scaffold that recruits histone modifying enzyme complexes, acting in a *trans* manner to do so.

More recently, a microarray containing more than 9400 human proteins was used as a platform to identify binding interactions between proteins and ten full-length coding and non-coding RNAs (Siprashvili *et al.*, 2012). The 137 protein–RNA interactions discovered were enriched for known human RNA-binding domains, including RRM, RBD, K homology domains, and zinc finger motifs. One such protein, Staufen 1 (Stau1), was identified as a binding target of *TP53* mRNA. Using *in vivo* RN pull-down assays, they were able to show that Stau1 bound sense *TP53* mRNA, but not control mRNA. This was validated using reciprocal pull-down experiments. Further characterization showed that Stau1 binding to *TP53* RNA aided in maintaining its RNA levels during transcriptional blockades.

Protein–small molecule interactions

Protein microarrays have also been shown to be useful in discovering new drug molecules and drug targets. Rapamycin is a small molecule drug that can induce a starvation response and inhibit cell growth through its target TOR (target of rapamycin), a highly conserved protein kinase regulating cell proliferation and metabolism, both in yeast and in humans. Huang *et al.* (2004) identified small-molecule inhibitors of rapamycin (SMIRs) and small molecule enhancers of rapamycin (SMERs). They obtained the binding partners of two SMIRs, SMIR3 and SMIR4 by probing the entire yeast proteome with biotinylated versions of the SMIRs, and two targets for rapamycin inhibition were identified. The first, Tep1p, is a homologue of the mammalian PTEN tumour suppressor; the second is Ybr077cp (Nir1p), a protein previously with no known function. Both Nir1p and Tep1p were shown to associate with PI(3,4)P2, which suggests a novel mechanism by which phosphatidylinositides might modulate targets of the rapamycin pathway and the ability of protein microarrays to identify specific protein–small molecule interactions.

Polyanions are a particular class of proteins with considerable negative charge and are hypothesized to be involved in intracellular organization, protein stabilization and folding, and protein transport. Salamat-Miller *et al.* (2006) used the yeast proteome array to probe for binding partners of well-characterized polyanionic proteins: actin, tubulin, heparin, heparan sulfate and DNA. By using biotinylated versions of these five polyanions, 893 polyanion-binding proteins (PABPs) were identified. The polyanions and their binding partners were shown to form a network involved with maintaining the structure and activity of yeast cells.

Protein–glycan interactions

One of the integral post-translational modifications involved in cellular membrane formation is glycosylation. Proper glycosylation is critical in dictating proper conformation of many membrane proteins, retaining stability of some secreted glycoproteins, and facilitating cell–cell adhesion. The yeast proteome array was used to further examine the roles of protein glycosylation in yeast (Kung *et al.*, 2009). Since the proteins were all purified in their native host, they were expected to maintain most of their PTMs. To that end, the arrays were probed with two fluorescently labelled lectins: concanavalin A (ConA), which recognizes mannose, and wheat-germ agglutinin (WGA), which recognizes N-acetylglucosamine (GlcNAc). Two separate yeast proteome arrays were used in this experiment. The first was the aforementioned array used in many of the previously discussed studies developed by Zhu *et al.* (2001), which used a GST affinity tag on the N-terminal and contained ≈ 5800 yeast proteins. The second array contained ≈ 5600 yeast proteins with a C-terminal Protein-A IgG-binding domain and was developed by Gelperin and colleagues (Gelperin *et al.*, 2005). There is an important distinction between these two arrays, especially as it pertains to glycoprotein signalling. Proteins that are anchored in the membrane at the C-terminal (and some type II proteins) are optimal for the N-terminal tagged array (Zhu) because the signalling sequence and transmembrane domains are present at the C-terminal and are unlikely to be effected by an N-terminal tag (e.g. cytochrome *b5* and the SNARE proteins). In contrast, proteome chips with C-terminal protein fusions (Gelperin) are optimal for studying type-I, and most of type-II and type-III membrane proteins that are glycosylated through the secretory pathway; the signalling and modifications with N-linked glycans occur at the N-terminal and would be hindered by an affinity tag at the N-terminal. ConA and WGA identified 124 and 174 proteins, respectively, in the N-linked array and ConA and WGA identified 236 and 142 proteins, respectively, in the C-linked array. In total, 534 proteins were identified, 406 of which were previously not known to be glycosylated. Gene ontology analysis identified enrichment in proteins from the secretory pathway, TFs, and other mitochondrial proteins. These mitochondrial protein targets were further examined by treating yeast cell cultures with tunicamycin, an inhibitor of N-linked protein glycosylation. Two of the four mitochondrial proteins identified showed partial distribution to the cytosol and reduced localization to the mitochondria, suggesting a new role of protein glycosylation in mitochondrial protein function and localization.

Another approach to identifying protein–glycan interactions involves the inverse layout of the microarray. A high-content lectin microarray consisting of 94 unique commercial lectins was fabricated and used to profile accessible surface glycans of mammalian cells (Tao *et al.*, 2008). Twenty-four human cell lines were labelled and applied to the microarray. Each cell line was subjected to a binary algorithm that was developed to generate 'glycan signatures', resulting in hierarchical clusters based on each line's accessible glycan composition. By comparing the glycan profiles of a breast cancer cell line and its cancer stem-like cell derivatives, three lectins (LEL, AAL, and WGA) were found to specifically recognize MCF7 cells, but not the derivatives. To further validate this results, the authors employed LEL-conjugated beads to purify away the normal MCF7 cells from the cancer stem-like cells (estimated as ≈ 0.1% in the cell population) in order to enrich for cancer stem-like cells. Next, by using a mouse model to test the enrichment of the cancer stem-like cells, they were able to show that two weeks following injection of the LEL-depleted cancer stem-like cell enriched cultures, the average tumour size was greater than 2-fold bigger than the control group injected with a similar number of normal MCF7 cells. This study demonstrates the utility of using lectins on a microarray to identify novel cell surface markers on cancer stem-like cells, and the ability to enrich a sample for cancer stem-like cells.

The affinities for lectins are typically very low (K_d is in the range of 10^{-3}–10^{-6} M), which can pose some difficulties when performing a cell-binding assay because low-affinity interactions may be washed away from the immobilized lectins. This is especially true when dealing with live cells. To overcome these problems, several researchers have modified this technology to improve the strength of the binding interactions. One example involves the antibody-assisted lectin profiling (ALP) approach developed by Kuno *et al.* (2009) for detecting glycoproteins at low concentrations. This method was used to analyse the glycan structures of the platelet aggregating factor hPod, which has been proposed to enhance the metastatic potential of glioblastoma cells. The hPod protein complex was first enriched by immunoprecipitation, before incubation on a lectin microarray to identify its associated glycans. The additional modification of the platform involves the use of an evanescent-field activated fluorescence detection system, which allows for a label-free, real time detection system. An evanescent field is generated within 200 nm of the solid surface, rendering the background signal so low that washing steps are not necessary. Additional studies have shown that this system is by far the most sensitive detection system among lectin microarrays, reaching a reported detection limit in the 100 pM range (Uchiyama, 2006).

Finally, Li *et al.* (2011) reported a two-phase discovery and validation scheme improving the sensitivity of the lectin microarray in the study of prostate cancer biomarkers. The approach first involves pooling tissue samples from four groups in equal amounts (50 μg): normal, nonaggressive cancer, aggressive cancer, and metastatic cancer. The discovery phase extracts prostate-specific antigen (PSA) and membrane metallo-endopeptidase (MME) proteins from each tissue group using an anti-total PSA antibody and an anti-MME mAB, respectively. The immunoprecipitated PSA and MME proteins were incubated on a lectin microarray, followed by detection of PSA and MME proteins using anti-PSA and anti-MME mAbs, respectively. Comparison of signals between each group of pooled tissue revealed that the fraction of PSA that is O-glycosylated (as recognized by jacalin) or Neu5Ac-conjugated (as recognized by SNA) was highly elevated in aggressive prostate cancer and metastatic prostate cancer groups. It was shown also that the fraction of MME that was modified by either GalNAx or GlcNac showed this elevation as well. Confirmation of these finding were shown using an immunosorbent assay, in which PSA and MME were first captured on an ECL plate coated with anti-PSA and anti-MME mAbs, followed by detection with biotinylated lectins. These studies show the power and adaptability of protein microarrays, even as they involve the detection of comparably weak protein–glycan or protein lectin interactions.

Detection of post-translational modifications

Protein post-translational modifications (PTMs) are one of the most important mechanisms used by cells to directly regulate protein activity. Hundreds of PTMs have been identified, and are enzyme-dependent and reversible, including protein (de)phosphorylation, (de) ubiquitylation, (de)SUMOylation, (de)acetylation, (de)nitrosylation, and (de)glycosylation. Many labs have attempted to further understand the biological consequences of these PTMs, as it is important to identify downstream targets at a systems level. 'Shotgun' MS/MS techniques have been a workhorse in the identification of PTMs in mammalian proteomes. However, this bottom-up approach does not identify the upstream enzymes responsible. The functional protein microarray provides a fantastic platform for the investigation of the players responsible for this complex category of cell signalling (Table 5.4).

Table 5.4 Summary of post-translational modification (PTM) studies using functional protein microarrays

PTM studied	Substrate	Enzyme	Reference
Phosphorylation	Yeast proteome	Yeast kinases	Ptacek *et al*. (2005), Zhu *et al*. (2000)
	3000 human proteins	Human CDK5	Schnack *et al*. (2008)
	Human TF array	Human kinases	Newman *et al*. (2013)
	Human proteome	Human CK2	Tarrant *et al*. (2012)
Ubiquitylation	Yeast proteome	Ubiquitin E3 Rsp5	Lu *et al*. (2008)
SUMOylation	Human	E3 RanGAP1	Oh *et al*. (2007)
Acetylation	Yeast proteome	NuA4 complex	Lin *et al*. (2009)
	E. coli proteome	PAT	Thao *et al*. (2010)
S-nitrosylation	Yeast proteome	N/A	Foster *et al*. (2009)
O-glycosylation	Human kinases	OGT	Dias *et al*. (2012)

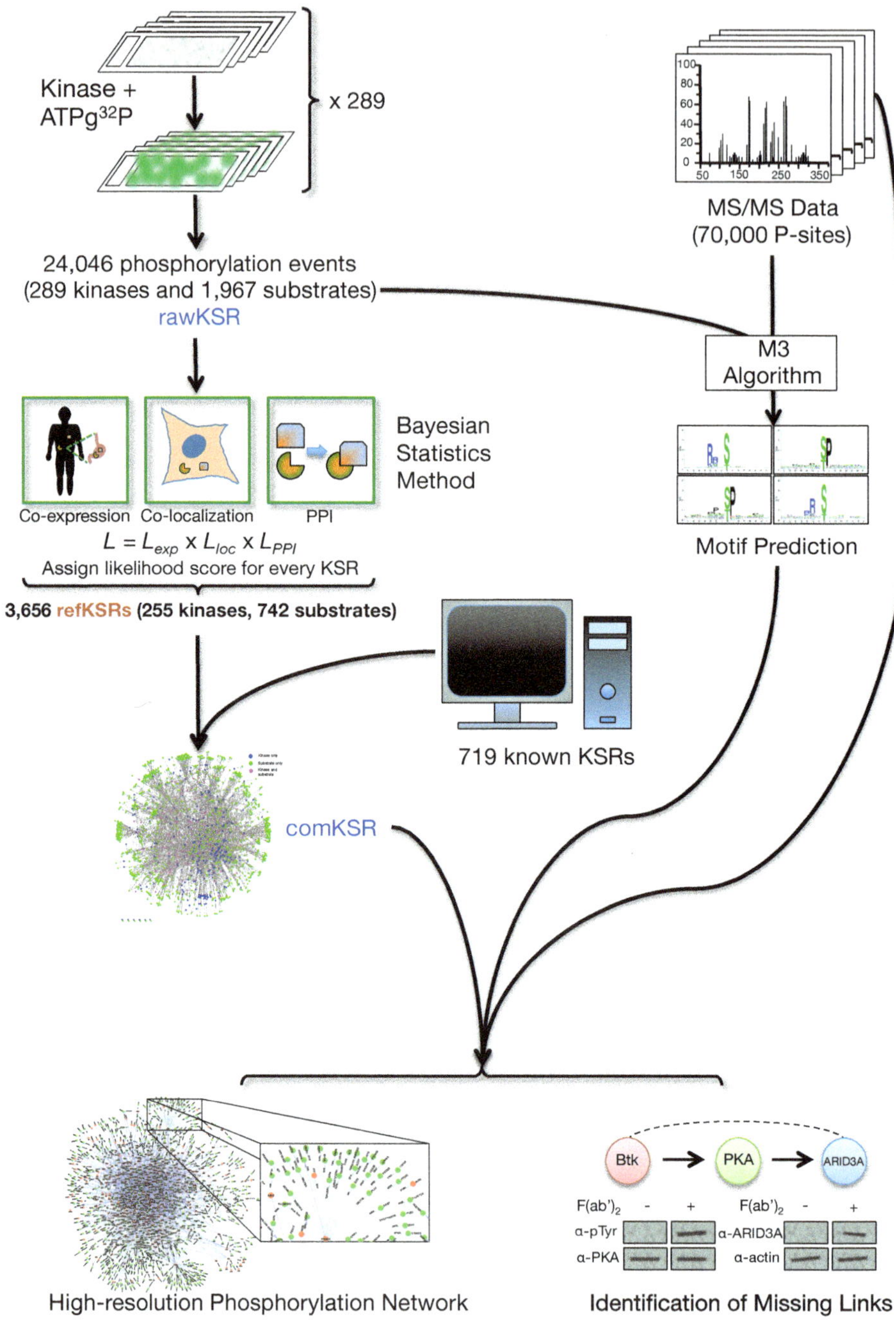

Figure 5.6 Schematic of the CEASAR strategy (connecting enzymes and substrates at amino acid resolution). 289 human kinases were used to treat protein microarrays containing 4200 human proteins, identifying over 24,000 phosphorylation events. Combining these data with a database of known phosphorylation sites and using the M3 algorithm led to the prediction of motifs for 284 kinases, which represents over 50% of the human kinome. In order to enrich the dataset for more physiologically relevant hits, a Bayesian statistics method was employed, which positively correlated hits that were co-expressed, co-localized, and were known to

Protein phosphorylation

Protein phosphorylation is one of the better-studied PTMs, and as such is considered one of the central players in most, if not all, cellular processes. The application of protein microarrays to the study of phosphorylation was first demonstrated by Zhu *et al.* in 2000. Seventeen different substrates were immobilized on a nanowell protein microarray, followed by individual kinase assays with 119 out of the 122 known yeast kinases (Zhu *et al.*, 2000). The *in vitro* phosphorylation was detected by using ^{33}P-γ-ATP as the phosphate donor in the kinase assay, and signal was acquired with either film or a phosphorimager screen, which can be used to quantify the isotopic signal. By using this approach, they were able to determine the kinase substrate specificity of the yeast kinome and identify a large number of tyrosine kinases.

As a follow-up to this project, the Snyder group performed a large-scale 'Phosphorylome Project' using the aforementioned yeast proteome arrays (Ptacek *et al.*, 2005). Eighty-seven yeast kinases were purified and incubated with the yeast proteome microarrays in the presence of ^{33}P-γ-ATP. A total of 1325 distinct protein substrates were identified, representing a total of 4129 phosphorylation events. Some observations were made immediately regarding the global kinase activity of yeast. Most (73%) substrates were recognized by fewer than three kinases, signifying the strong specificity of kinases for their substrates. Transcription factors were the largest class of proteins that were phosphorylated, supporting the regulatory role of phosphorylation on protein expression. Further analysis provided a global network that connected kinases to their potential substrates, offering new opportunities to identify new signalling pathways and crosstalk between pathways.

The natural evolution of this research reached higher eukaryotes in early 2013 when Newman *et al.* (2013) analysed human kinase–substrate relationships (KSRs) using the TF microarray developed in the laboratory previously (Hu *et al.*, 2009) which contained 4191 unique, full-length human proteins. To date, approximately 2000 human KSRs have been experimentally verified, i.e. a known kinase has been shown to phosphorylate a known substrate. This pales in comparison to the over 70,000 phosphorylated serine, threonine, and tyrosine residues that have been identified *in vivo* through mass spectrometry (MS/MS). However, as mentioned before, simply knowing which residues are modified does not give any insight to the pathways or even the direct upstream kinase that is responsible for the modification. A new strategy, dubbed CEASAR (connecting enzymes and substrates at amino acid resolution), was employed to fabricate a high-resolution map of human phosphorylation networks (Fig. 5.6), connecting kinases to their immediate downstream targets (Newman *et al.*, 2013). The TF microarrays were incubated with 289 unique, active, and full-length human kinases in the presence of ^{32}P-γ-ATP. These experiments yielded 24,046 phosphorylation events involving all 289 kinases and 1967 unique substrates to create a 'rawKSR' dataset. Using a Bayesian

---

interact in literature. This, combined with a dataset of known kinase–substrate relationships (KSRs), formed a common dataset, which when combined with the motif predictions and the MS/MS data allowed for the generation of a high-resolution phosphorylation network. This network shows not only which kinase phosphorylates which substrate, but also predicts the site(s) of phosphorylation based on the motif predictions. A missing link was also identified in the Btk pathway, strengthening not only the bioinformatics used, but also the overall validity of this technique.

statistics model, the raw KSR dataset was refined by hypothesizing that the KSRs would be more physiologically relevant if they were known to share the same tissue specificity and sub-cellular localization, and if they were already known to interact from literature. A high-confidence dataset was created containing 3656 refined KSRs (refKSRs) involving 255 unique kinases and 742 substrate proteins. Finally, a combined (comKSR) dataset was created that included both the refKSR data and 719 known KSRs. Validation studies in HeLa cell lines showed that 76% of tested KSRs showed some sort of kinase-dependent change in the substrate (e.g. stabilization, degradation, or mobility shift). At the same time, an integrated algorithm was developed, termed M3 (Motif discovery based on Microarray and MS/MS), to systematically identify phosphorylation motifs. Combining the large MS/MS datasets with the rawKSR and known KSR dataset, over 13,000 of the 70,000 known phospho-sites identified by MS/MS were mapped to 1644 substrates. Three hundred consensus motifs for 284 kinases were identified, representing 55% of the human kinome. Integrating the information from both the motif dataset and the *in vivo* phosphorylation site dataset into the comKSR dataset created a high-resolution phosphorylation map connecting 230 kinases to 2591 *in vivo* phosphorylation sites on 652 substrates, and identifying over 3600 new KSRs. Using the newly generated phosphorylome map allowed an intermediate kinase to be identified in BCR signalling. Protein kinase A (PKA) was identified as both a substrate for Bruton's tyrosine kinase (BTK) and a kinase for ARID3A, filling in a gap that had not been identified before. The phosphorylations of both PKA and ARID3A were validated *in vivo*, and two phosphorylation sites on ARID3A that were predicted using M3 were verified. It was shown that BTK phosphorylates and activates PKA during BCR signalling, which in turn leads to the stabilization of ARID3A upon phosphorylation by PKA. Taken as a whole, this study represents a bright future of what functional protein microarrays can provide – a global understanding of cell signalling pathways through careful experimentation and bioinformatics.

Several smaller-scale studies of kinase–substrate interactions have been reported as well. For instance, Popescu *et al.* (2009) probed an array containing 2158 *Arabidopsis* proteins with 10 *Arabidopsis* mitogen-activated protein kinases (MPKs). They identified 570 putative MPK phosphorylation targets, which were enriched in transcription factors involved in the regulation of development, defence, and stress response. A commercially available human protein microarray comprised of approximately 3000 individual proteins was used to identify substrates of cyclin-dependent kinase 5 (Cdk5), a serine/threonine kinase that plays an important role during central nervous system development (Schnack *et al.*, 2008).

As shown previously in the protein–RNA interaction applications, viral proteome microarrays are a very useful tool in the study of host–pathogen interactions. In 2009, Zhu and colleagues used an Epstein–Barr herpes virus (EBV) protein microarray to investigate the function of an EBV-encoded protein kinase, BGLF4, via phosphorylation and binding assays (Zhu *et al.*, 2009). They identified a total of 23 BGLF4 substrates and interactors, including EBNA1, a protein that is essential for the replication and maintenance of the episomal EBV genome during latency. The authors were able to show that BGLF4 acts as a negative regulator of EBNA1's replication function and raised the possibility that the induction of BGLF4 kinase activity may provide a novel means of eliminating EBV genomes from latency in infected cells.

Protein ubiquitylation

Ubiquitylation is one of the most prevalent PTMs in eukaryotes and controls a variety of intracellular signalling events, but its regulatory mechanisms are largely unknown. The use of protein microarrays represents a rapid and high-throughput method of identifying players in this pathway, and Gupta and colleagues did just that in 2007 (Gupta *et al.*, 2007). Rsp5 is a well characterized HECT-domain E3 ligase from yeast belonging to the Nedd4 family. Using a commercial yeast proteome array for duplicate reactions, they were able to identify 40 high-confidence Rsp5 substrates. Rsp5 contains five WW domains, which are known to bind to specific substrates by recognizing a (L/P)PXY sequence known as a PY motif. Not surprisingly, 72% of these substrates contained at least one of these sequences. Many of these were further validated *in vitro* and *in vivo*, indicating the fidelity of this assay. After identification of the substrates of Rsp5, they again turned to the microarray to identify its binding partners by treatment with a fluorescently labelled Rsp5, which again showed enrichment of proteins containing a PY domain. From this, they were able to build both an interaction network and a substrate network.

At the same time, Lu *et al.* (2008) used a protein microarray approach for the identification of Rsp5 substrates as well, but were able to further characterize two of the substrates. Using the Zhu lab yeast proteome array, more than 90 new substrates were identified, and eight were validated as *in vivo* substrates of Rsp5. Characterization of one substrate, Rnr2 (a ribonucleotide reductase), revealed that Rsp5-dependent ubiquitylation affects subcellular localization. Heterozygous null mutants of Rnr2 correspond to hypersensitivity to DNA damage and treatment with the RNR specific inhibitor, hydroxyurea (HU). After DNA damage, another member of the RNR family, Rnr4, is redistributed within cells, but it is not known whether it is due to a PTM. They were able to show that Rnr2's localization was dependent on Rsp5, as it was localized to both the cystosol and nucleus in a *wt* strain, but only localized to the nucleus in an Rsp5 knockout strain. Both of these previous studies highlight the power that protein microarrays can hold in the analysis of ubiquitylation.

While each of the previous assays described were able to identify true *in vivo* interactions in an *in vitro* setting, one limitation is apparent, as none can mimic the precise cellular conditions used by these proteins *in vivo*. This can lead to false positives due to the promiscuity of the applied protein, or false negatives due to some of the necessary cofactors being absent from the reaction mix. The use of concentrated mammalian cell extracts in combination with protein microarrays can serve to identify PTM targets in a semi-*in vivo* setting while alleviating the challenge of analysing a complex mixture. Merbl and Kirschner generated from HeLa S3 cells three cell extracts from the two distinct phases surrounding anaphase: the mitotic checkpoint (treated with nocodazole, which prevents mitotic spindle formation and checkpoint arrest), the anaphase release phase (treated with the E2 ligase UbcH10), and an anaphase release phase containing an APC (anaphase promoting complex) inhibitor as well as UbcH10 (Merbl and Kirschner, 2009). The synchronized cell extracts were incubated with Invitrogen's Human ProtoArray (composed of 8000 human proteins) and using the anti-polyubiquitin antibody FK1, identified 132 polyubiquitylated proteins. The authors expected to recover substrates of APC and were able to identify 11 of the 16 known APC targets on the array, confirming the experimental design. Validation studies performed in rabbit reticulocyte lysate confirmed the degradation/ubiquitylation of seven novel APC substrates. This study demonstrates the efficacy of using protein microarrays in combination with cell extracts to recapitulate the global PTM signature in a specific cellular state.

Protein acetylation

Acetylation of histone residues by histone acetyltransferases (HATs) and deacetylation by histone deacetylases (HDACs) has been shown to be a key regulator of chromatin structure and transcription. While histones appear to be the dominant substrate for this modification, it has been hypothesized that many other non-histone proteins may be modified by these enzymes. For example, the HAT Esa1, which is part of the essential nucleosome acetyltransferase of H4 (NuA4) complex, is the only vital HAT in yeast, suggesting that it may target other non-histone proteins critical for survival. Non-histone substrates of the NuA4 complex were identified by Lin *et al.* (2009) by performing acetylation reactions on the yeast proteome microarrays. Arrays were treated with purified NuA4 complex in the presence of ^{14}C-acetyl-CoA as the donor, and 91 proteins were found to be readily acetylated. Further validation of 20 targets revealed 13 with Esa1-dependent acetylation in cells. Phosphoenolpyruvate carboxykinase (Pck1p) was further characterized by mass spectroscopy, where acetylation was observed on Lys19 and Lys514. Mutagenesis at these sites demonstrated that K514 acetylation is critical in enhancing the activity of Pck1p, resulting in a longer lifespan for yeast growing under starvation.

S-Nitrosylation

S-Nitrosylation is a unique PTM in that it is independent of enzyme catalysis, but is nonetheless an important one that affects a wide range of protein and cellular processes, including inflammation and protection from apoptosis. Recently, Foster *et al.* (2009) developed a protein microarray-based approach to identify proteins reactive to S-nitrosothiol (SNO), the donor of NO^+ in S-nitrosylation. S-Nitrosocysteine (CysNO), a highly reactive SNO, was added to a commercially available yeast microarray. Nitrosylated proteins were detected using a modified biotin switch technique (BST), which converts an S-nitrosothiol into an S-biotinylated Cys. The arrays were subsequently probed with an anti-biotin antibody, followed by a fluorescently labelled secondary antibody. The top 300 proteins with the strongest signal were further characterized and showed that proteins with active site Cys thiols residing at N termini of alpha helices or within catalytic loops were particularly enriched. However, even among these protein families, there was significant variation in the S-nitrosylation, signifying that secondary structure or intrinsic nucleophilicity of Cys thiols was not sufficient to interpret the specificity of S-nitrosylation. Further analyses revealed that NO-donor stereochemistry and structure had significant impact on S-nitrosylation efficiency.

O-GlcNAcylation

Although many, if not most, proteins have been identified as phosphorylated on serine and threonine residues, modification by β-N-acetylglucosamine (O-GlcNAc) on serine and threonine residues has emerged as another fundamental regulatory mechanism in cell signalling. O-GlcNAc cycling on proteins is mediated by two enzymes: O-glycosyltransferase (OGT), which glycosylates S/T residues, and O-GlcNAcase (OGA), which hydrolyses S/T residues back to their native state. Like phosphorylation, O-glycosylation can modify protein function, including kinase activity, turnover, protein–protein interactions, subcellular localization, DNA affinity, and transcription activity. Dias *et al.* (2012) were the first to use a functional protein microarray to probe for O-GlcNAc on a large scale. A commercially available kinase array was used which contained 152 full-length human kinases. Using tritiated

UDP-GlcNAc (UDP-^{3}H-GlcNAc) as the donor along with purified OGT, they were able to identify 42 kinases modified *in vitro*. Further validations in HEK293 cells were performed that confirmed the modification occurs *in vivo* as well. While phosphorylation is controlled in humans by over 500 kinases and over 150 phosphatases, there is only one OGT and one OGA, suggesting that the specificity with which these enzymes react most likely involves other binding partners, and further studies will need to be performed in order to elucidate the breadth of this modification.

Profiling monoclonal antibody specificity

Antibodies have widespread applications in proteomic studies, but some difficulty lies in producing antibodies with sufficient specificity. Monoclonal antibodies (mAbs) are a better option than polyclonal antibodies for most applications, but tend to be much more expensive and can sometimes be too specific. Protein microarrays offer a technique to be able to test the specificity of these antibodies. After immunizing mice with live cells from human livers, Hu *et al.* (2007) isolated 54 hybridomas with binding activities to human cells and identified antigens for five mAbs by screening on a protein microarray consisting of 1058 unique human liver proteins. The five identified mAbs were subsequently used to characterize the expression profiles of their corresponding antigens in both normal liver cells and hepatoma cells. Among them, eIF1A, an essential initiation factor in translation, was found to be present in normal hepatocytes, but not in any hepatoma cells, suggesting that liver carcinomas likely have suppressed translation.

In 2012, Jeong and colleagues used a similar approach combining immunization with live human cells and even larger microarray-based analysis to rapidly identify monospecific monoclonal antibodies (mmAbs) (Jeong *et al.*, 2012). The protein microarray used in this study is one of the largest to date, as it contains almost 17,000 individually purified full-length human proteins. When the monoclonal antibodies were used on this array and only recognized a single antigen, they could be identified as monospecific mAbs. These antibodies were tested against their antigen and could be successfully identified in Western blots (WB), successfully immunoprecipitate endogenous antigen from homogenate, and successfully used in immunocytochemistry (ICC) in transfected human cell lines. The specificity with which this approach created mmAbs was more useful (e.g. more applications – WB, IP, ICC/IHC) than many other commercially available sources. For example, 66% of the mmAbs tested were IP-grade, compared to the commercial average of 18%. Future studies may be able to use this approach to tackle the issues analytical microarrays possess by easily and rapidly producing highly specific mmAbs.

Clinical research applications

As the basic research field continues to profit from protein microarrays, so does the clinical field. The ability to more sensitively and rapidly screen patient samples against a large array of proteins inches us closer to what many consider the future of health care – personalized medicine. While we certainly are not at the level yet, there is great promise shown in many of the studies that have arisen the last decade. In this section, we will discuss three applications of protein microarrays in the clinical field: biomarker identification, pathogen–host interactions, and cancer research (Table 5.5).

Table 5.5 Summary of clinical disease studies using protein microarrays

Disease type	Disease	Substrate	Reference
Infectious	SARS infection	82 viral proteins	Zhu *et al.* (2006)
	B-cell lymphoma/AIDs-related Kaposi's lymphoma	174 EBV and KSHV viral proteins	Zheng *et al.* (2011)
	Rabbit model of the plague	149 proteins from *Yersinia pestis*	Li *et al.* (2005)
	Brucellosis	3046 proteins expressed in lysates	Liang *et al.* (2011)
	Cervical carcinomas	154 proteins from 13 papillomavirus	Luevano *et al.* (2010)
	Streptococcus infection	201 viral proteins from 2 strains	Margarit *et al.* (2009)
Autoimmune	Inflammatory bowel disease	*E. coli* arrays with 4179 proteins	Chen *et al.* (2009)
	Autoimmune hepatitis	5011 human proteins	Song *et al.* (2010)
	Primary biliary cirrhosis	17,000 human proteins	Hu *et al.* (2012)
	Sjögren's syndrome	8000 human proteins	Hu *et al.* (2011)

Biomarker identification

One of the most rapidly growing applications in the field of clinical proteomics using protein microarrays is biomarker identification. This was first used in traditional serology studies, which focused on diagnostic identification of antibodies in patient serum samples. These antibodies are produced as a part of the immune response to an infection, a foreign protein, or, in the case of autoimmune disease, against ones own proteins. Using protein microarrays as a platform for potential antigens, researchers can identify autoantibodies with statistical significance and association with an infection or disease of interest. Generally, the patient sera are first diluted (e.g. 1000-fold) before incubation on a pre-blocked antigen microarray (e.g. protein microarray) and ending with a stringent washing step. Then, positive signals are detected using anti-human IgG, IgM, or IgA antibodies conjugated to various fluorophores. Compared to traditional serology techniques, such as ELISA, agglutination, precipitation, complement fixation and fluorescent antibodies, protein microarray-based profiling is much more sensitive, unbiased, and can be performed at a much higher throughput. This next section will review four studies illustrating the history and development of protein microarrays in biomarker identification.

SARS-CoV diagnosis

The first viral proteome microarray was fabricated by Zhu *et al.* (2006) and consisted of every full-length protein and protein fragment encoded by the SARS coronavirus (SARS-CoV) as well as proteins from five additional mammalian coronaviruses. These microarrays were used to screen 400 Canadian serum samples that were collected during the 2002 SARS outbreak. Included samples were those confirmed as SARS-CoV positive, other respiratory illness patients, and healthcare professionals. Antibody response was quantified by using both human IgM and IgG antibodies coupled to different fluorophores. In order to identify biomarkers, the serum samples were first clustered according to the relative signal intensities of all coronavirus proteins in an unsupervised fashion. Two major groups were identified, which, when compared to clinical data, were largely correlated with either SARS-positive or

SARS-negative sera. Five fragments of the SARS nucleocapsid protein (N protein) associated tightly with SARS infection, as well as one spike protein (S protein) fragment. However, a few proteins encoded by other coronaviruses also displayed significant correlation. In order to determine the best classifiers and classification model, two different supervised analysis approaches were applied; k nearest neighbour (k-NN), which measures the similarity between a new case and all known cases, and logistics regression (LR), a generalized linear regression for binary response. The N protein of SARS-CoV and the S protein from both SARS-CoV and HCoV-229E were identified as the best classifiers. One useful feature of a serum test over a nucleic acid diagnostic test is that anti-pathogen antibodies can potentially be detected long after infection. To that end, serum samples collected from SARS patients who recovered from respiratory disease (≈ 320 days after diagnosis) were used to probe the microarray, where positive signals were detected with both anti-human IgM and IgG antibodies. The results showed that SARS N proteins could be readily recognized by human IgG and, importantly, not by IgM antibodies. However, serum samples collected from Chinese patients immediately after fever was detected showed much stronger signal both in IgG and IgM profiling. These results show that protein microarrays can be used to detect both early response anti-pathogen antibodies as well as late response antibodies long after infection. This approach is potentially applicable to all viruses and is expected to have a large impact on both epidemiological studies and clinical diagnoses.

Humoral immune responses to herpes viruses

A similar approach has been used to profile humoral immune responses to two human herpes viruses, the Epstein–Barr virus (EBV) and Kaposi's sarcoma-associated herpesvirus (KSHV). While EBV is a ubiquitous human herpesvirus, KSHV has a much more restricted seroprevalence. Both viruses have been shown to be associated with malignancies and also display an increased frequency in individuals who are also infected with human immunodeficiency virus type 1 (HIV-1). To investigate humoral immune responses, a protein microarray consisting of 174 EBV and KSHV full-length proteins was generated by the Zhu and Hayward groups (Zhu *et al.*, 2009; Zheng *et al.*, 2011). Plasma antibody responses to EBV and KSHV were examined from healthy volunteers and patients with B-cell lymphoma, or with AIDS-related Kaposi's sarcoma or lymphoma. Apart from detecting IgG responses from known antigens, the tegument proteins ORF38 (KSHV), BBRF (EBV), BGLF2 (EBV), and BNRF1 (EBV), and the EBV early lytic proteins BRRF1 and BORF2, were also detected. IgA responses to EBV EBNA1 and viral capsid antigens have been used as a diagnostic tool for nasopharyngeal carcinoma for years, but the same IgA response was also found in healthy and HIV-infected patients. Comparing the IgG and IgA responses showed that IgA responses were much higher against BCRF1, BRRF2, and LMP2A. This study demonstrates that even plasma can be used for biomarker identification and that other immunoglobulin isotypes such as IgA are worth considering when studying immune response.

E. coli proteome microarrays for IBD diagnosis

Crohn's disease (CD) and ulcerative colitis (UC) are chronic, idiopathic and clinically heterogeneous intestinal disorders that are collectively known as inflammatory bowel disease (IBD). Although IBDs have been implicated in autoimmune disease, antibodies against microbes have been seen in the sera of IBD patients, some of which have been used as biomarkers for diagnosis and prognosis of the disease. Using the same *E. coli* K12 protein microarray that was used to study protein–DNA interactions mentioned earlier, Chen *et al.* (2009) decided to profile serum samples collected from CD and UC patients. The proteome array was screened using individual serum from healthy controls ($n = 39$) and clinically well characterized patients with CD ($n = 66$) and UC ($n = 29$). To their surprise, among the 417 *E. coli* proteins that were differentially recognized by serum antibodies from healthy controls and either IBD patient, 169 proteins were identified as highly immunogenic in healthy controls and 186 were identified as highly immunogenic in CD patients, but only 19 proteins were identified as highly immunogenic in UC patients. Through statistical analysis, they were able to identify two sets of serum antibodies as novel biomarkers for distinguishing CD from healthy controls (accuracy, 86 ± 4%; $P < 0.01$) and CD from UC (accuracy, 80 ± 2%; $P < 0.01$). This was the first demonstration of using proteome microarrays for the discovery of novel serological markers, as well as the first to examine human immune responses to the entire proteome of a microbial species in a disease context.

Autoantigen discovery for autoimmune hepatitis (AIH)

A microarray consisting of individually purified human proteins would seem an ideal tool for the discovery of new autoantigens associated with autoimmune disease. Autoimmune hepatitis (AIH) is a chronic necroinflammatory disease of the human liver with little known aetiology. Although the detection of non-organ-specific and liver-related autoantibodies using immunoserological approaches have been widely used for diagnosis and prognosis, these traditional autoantigens, such as anti-SMA (smooth muscle autoantibodies) and anti-ANA (antinuclear autoantibodies) are often mixtures of very complex biological materials. Identification and characterization of these autoantigens is dependent on their unambiguous and accurate detection, which is not possible using the traditional means. To address this, Song and colleagues created a human protein microarray of 5011 non-redundant proteins that were expressed and purified from yeast (Song *et al.*, 2010). In order to make this technique feasible in a clinical application, the cost of performing such an assay has to be considered, as a single human protein array with 9000 proteins can cost upwards of $1000. Therefore, a two-stage strategy was employed to limit costs in identifying new biomarkers in AIH (Fig. 5.7). Phase I consists of rapid selection of candidate biomarkers, which are then validated in a smaller array in phase II. In phase I, 30 AIH and 30 control serum samples were obtained and individually used to probe the human protein microarrays at a 1000-fold dilution. This was followed by detection using a Cy5-conjugated anti-human IgG antibody. After statistical analysis, 11 candidate autoantigens were found. In order to validate the candidates, the 11 proteins and three positive controls were purified again to use in the production of a large number of low-cost small arrays for Phase II validation. These new, smaller arrays were sequentially probed with the serum samples used in Phase I as well as from serum samples obtained from an additional group of patients consisting of 22 AIH, 50 primary biliary cirrhosis (PBC), 43 hepatitis B (HB), 41 hepatitis C (HC), 11 system lupus erythematosus (SLE), 11 primary Sjögren's syndrome (pSS), and two rheumatoid arthritis (RA) samples. Twenty-six serum samples from patients with other types of severe disease and 50 samples from healthy subjects were used as negative controls. From these Phase II studies, three new antigens, RPS20, Alb2-like, and dUTPase, were identified as highly AIH-specific biomarkers with sensitivities of 47.5%, 45.5%, and 22.7%, respectively. These were further validated with additional AIH samples in a double-blind design, and were also able

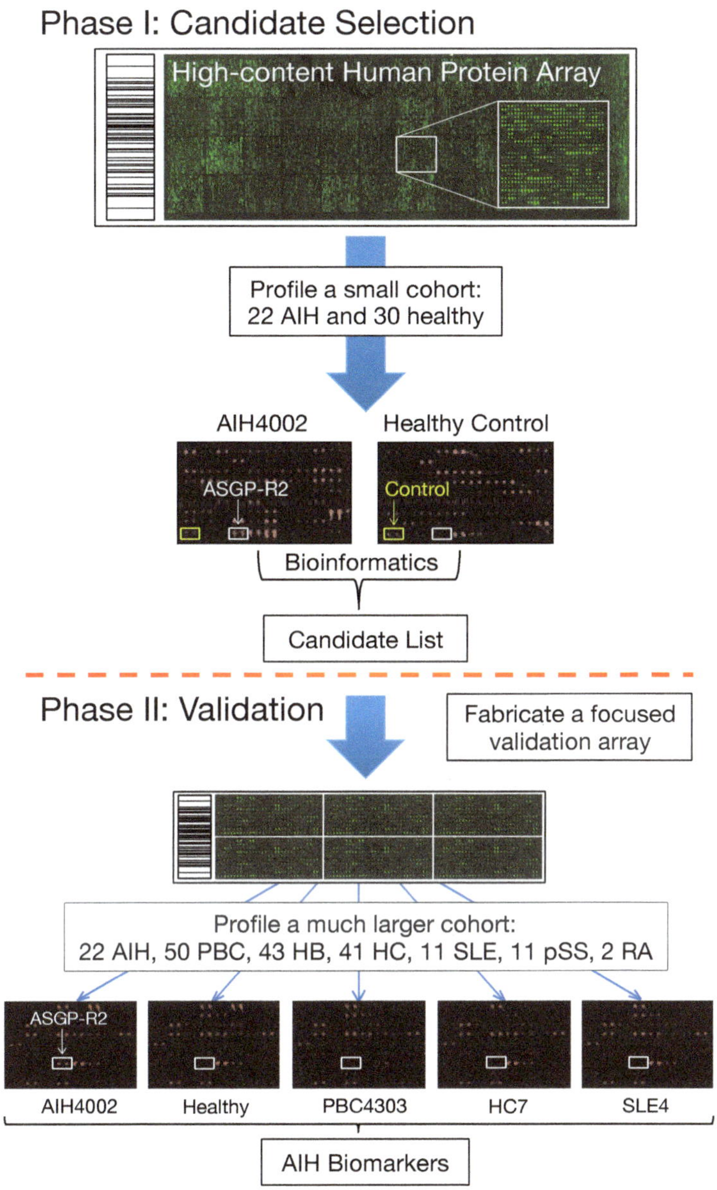

Figure 5.7 Schematic of the two-phase strategy used for biomarker identification in human autoimmune hepatitis (AIH). Phase I uses a small cohort to screen candidate biomarkers using serum profiling assays on high-density human protein microarrays. These microarrays are large and expensive, so using a small sample size reduces the cost considerably. Phase II uses the candidate list acquired from phase I to fabricate a smaller, more focused array in order to perform a much larger screen in a double blind manner. This approach represents a cost-effective way to identify biomarkers and is a scheme that should be replicated in future studies.

to demonstrate that these new biomarkers could be easily applied to canonical ELISA-based assays for clinical diagnosis and prognosis.

This study represents a novel paradigm in biomarker identification using protein microarrays for three reasons. First, a manageable number of candidate biomarkers can be quickly identified at a low cost because fewer expensive microarrays are required for phase I. Second, by using a smaller microarray consisting only of selected candidate proteins, the validation step can be quickly carried out in higher throughput with a much lower cost. This is the crucial step in the validation of these markers if one is to avoid an 'overfitting' problem that can arise when a candidate list is smaller than 40, which is likely the case for many of these applications. Overfitting becomes an issue when a statistical model describes random error instead of identifying a relationship. If a system is excessively complex, as is the case with most biomarker identification screens, then there are too many individual-to-individual variations relative to the number of samples used, leading to poor predictive performance. Therefore, testing in an additional, larger cohort using a double-blind protocol is an effective way to overcome this issue. Third, the author developed an ELISA-based assay to examine the performance of the newly identified biomarkers, serving as a translational step towards clinical practice.

Several other studies have employed pathogen protein microarrays to profile serological responses following infection as well, including bacterial and viral arrays used for biomarker identification in various infectious diseases (Liang *et al.*, 2011; Vigil *et al.*, 2011; Luevano *et al.*, 2010; Doolan *et al.*, 2008). While these studies have clearly demonstrated the power of protein microarrays in the identification of potential biomarkers, several shortcomings are repeatedly seen in these studies. For example, many of these arrays were fabricated using proteins translated in *E. coli* lysates without purification. Because these proteins are contaminated with unwanted *E. coli* proteins, the sensitivity of the assay is likely reduced due to their high immunogenicity, although using *E. coli* lysates to block the array alleviates this problem to a degree. More importantly, many of the biomarkers identified in these studies were not validated with additional cohorts, therefore, the possibility of overfitting is not completely ruled out.

Pathogen–host interactions

Another emerging application for protein microarrays in clinical proteomics is the unbiased, proteome-wide survey of important players involved in pathogen–host interactions. The identified factors, whether encoded by either a pathogen or a host, have the potential to be developed into potent therapeutic targets. This strategy is particularly useful for investigating virus–host interactions because after entering the host cells, the viral genome and its encoded proteins are in direct physical contact with the host's biological materials. This section will cover three ways in which these interactions can be investigated: RNA–protein interactions, enzyme–substrate relationships, and protein–protein interactions.

BMV RNA and host proteome interactions

The yeast proteome array was used to identify host factors that can affect the replication of Brome Mosaic Virus (BMV), an RNA virus that infects plants, but can also replicate inside of *S. cerevisiae* (Zhu *et al.*, 2007). Previous studies had shown that this positive-stranded RNA virus encodes a tRNA-like structure at the 3′-end of its RNA genome, where a clamped adenine motif (CAM) is required for the packaging of its genome into the capsid. To identify the necessary host proteins that could interfere with the viral packaging process, the yeast proteome arrays were incubated with a Cy3-labelled CAM-containing RNA stem–loop structure in the presence of a Cy5-labelled mutated CAM hairpin. By taking the

Cy3:Cy5 ratio, the top hits were identified and validated using an *in vitro* gel-shift assay. Two candidate proteins, pseudouridine synthase 4 (Pus4) and the actin patch protein 1 (App1), were further characterized in *Nicotiana benthamiana*. Both were shown to reduce BMV plus-strand RNA accumulation modestly, but had a much greater effect on the ability of BMV to spread in plants. Pus4 was also shown to prevent the encapsidation of the BMV RNA and reassembly of virions, providing a regulatory link between CAM-containing RNA motifs and encapsidation. This is yet another example of the considerable versatility of protein microarrays. Not only can they be used to study prokaryotic and eukaryotic interactions, but viral interactions as well, in an unbiased and systematic manner.

Host phosphorylome of virus-encoded kinases

Viruses have been very successful in exploiting their hosts in developing their own 'arsenal,' sometime using both host DNA and proteins in the process. Understanding the method by which the virus interacts with the host machinery is critical in developing effective antivirals. The human α-, β- and γ-herpesviruses differently infect tissues and cause distinct diseases, ranging from mild cold sores to pneumonitis, birth defects, and even cancers. However, they each confront similar challenges in infecting their hosts, namely reprogramming cellular gene expression, sensing and modifying cell-cycle progression, and reactivating the lytic life cycle to produce new virions and spread infection. Across the herpes virus family, many orthologous serine/threonine kinases are conserved that are involved in replication of the viral genome during the lytic life cycle. Therefore, many have postulated that if there are any shared substrates between these orthologous viral kinases, the targeted host proteins critical for viral replication would be revealed and aid in distinct and effective antiviral targets.

In testing this hypothesis, Li and colleagues purified four orthologous kinases encoded by EBV, KSHV, HCMV (human cytomegalovirus), and HSV-1 (herpes simplex virus 1) and incubated them on a human protein microarray (Li *et al.*, 2011). They identified 110 shared substrates and applied Gene Ontology (GO) and STRING analyses (a database of known and predicted protein–protein interactions – http://string-db.org/) to these candidates, finding a highly connected cluster of 15 proteins. Surprisingly, these proteins were all shown to be associated with the DNA damage response (DDR) pathway. The host DDR has been known to be an important player for many viruses, including human herpes viruses, and is also relevant to virus-induced tumorigenesis (Nikitin *et al.*, 2012). To further narrow down the potential players in this pathway, it was reasoned that an upstream master regulator of the pathway would be a likely target. After literature searches, Tat-interactive protein 60 (TIP60) emerged as an excellent candidate for four reasons: (1) TIP60 is further upstream in the DDR pathway than the other candidates; (2) it serves as a master regulator in DDR by acetylating ATM and initiating its autophosphorylation activity; (3) it is a histone acetyltransferase, and therefore can regulate chromatin dynamics through histone acetylation; and (4) it has been implicated in other viruses. When TIP60 was knocked down in EBV-infected B-cells, EBV's lytic replication was greatly reduced. Next, the authors performed a series of cell-based assays showing that during EBV replication, TIP60 activation by the BGLF4 kinase triggers EBV-induced DDR and mediates induction of viral lytic gene expression. Efficient lytic replication was also shown to be dependent on TIP60 in KSHV, HCMV, and HSV-1.

This work illustrates that high-throughput, unbiased approaches for the identification of conserved viral host targets show great potential for novel therapeutic antiviral targets.

Treatment of herpes virus is limited to very few drugs due to viral escape mutants that arise from extensive use, rendering these kinases attractive targets for future studies. Like all drug studies, development of effective drugs requires extensive knowledge of cellular targets. This novel 'common substrate identification' approach using microarrays can aid in the design of new assays for new and broadly effective anti-herpesvirus therapeutics.

LANA interactome analysis reveals a role in telomere shortening

While we have already discussed using protein microarrays to measure protein–protein interactions, it can also be specifically used to profile interactomes between a pathogen and a host. Recently, Shamay and colleagues examined interactions between the KSHV-encoded virulent factor, LANA, and a human host using the transcription factor microarray described before (Shamay *et al.*, 2012). LANA is an essential participant in KSHV genome replication and dysregulated cell growth in latently infected cells. Its interactions with host proteins have been extensively studied using a variety of techniques (including Y2H, GST affinity, IP assays, and HPLC-MS/MS), but each approach identified a different group of proteins. This study used a FLAG-tagged LANA to identify 61 potential binding partners, many of which were novel. Further validations confirmed eight of nine by co-immunoprecipitation and included TIP60, protein phosphatase 2A (PP2A), replication protein A (RPA) and the DNA-repair protein XPA. Although human papillomavirus (HPV) E6, HIV-1 TAT, and HCMV pUL27 interact with TIP60 to induce its degradation, LANA-associated TIP60 retained its HAT activity and showed increased stability. This finding mirrors what was described in the previous section that showed TIP60 plays a positive role in KSHV lytic replication. The identification of LANA as an RPA binding partner suggests that it may also have a role in the replication of cellular telomeric DNA. To test this, the authors performed ChIP assays using anti-RPA1 and –RPA2 antibodies and primers specific for the telomeric regions of DNA. They found that when LANA was present, RPA1 and RPA2 recruitment to telomeres was inhibited, but had no impact on the protein levels of the RPA complex, which implicated LANA as a factor in telomere length. Using Southern blots to analyse terminal restriction fragments showed that the length of telomeres was shortened by at least 50% in both LANA-expressing endothelial cells and KSHV-infected primary effusion lymphoma cells. Although it was shown that LANA effects telomere length, many more questions regarding its role need to be answered to fully understand how it is used by KSHV.

SUMO-EBV interactome reveals a new mechanism of EBV lytic replication

Inversely, a human factor of interest can be used to probe a virus protein microarray to identify crucial viral factors. The small ubiquitin-related modifier (SUMO) pathway is similar to the ubiquitin pathway because it involves a series of sequential enzymatic reactions to conjugate SUMO to lysine residues on substrate proteins. It has been shown that both latent and lytic EBV proteins can interact with the SUMOylation pathway, with non-covalent interactions occurring between SUMO and EBV proteins via a SUMO interaction motif (SIM). In order to systematically identify EBV proteins that bind to SUMO, Li *et al.* (2012) performed binding assays with human SUMO2 using the previously described EBV proteome microarray. Eleven proteins were identified, including the conserved viral kinase BGLF4. Mutations at putative SIMs within BGLF4 at both the N- and C-termini changed the intracellular localization from the nucleus to the cytoplasm. If mutations were made at only the N-termini, localization remained mostly in the nucleus, and localization was in both the cytoplasm and the nucleus when mutations were made at only the C-termini. BGLF4 was shown to inhibit the EBV lytic cycle transactivator ZTA by abolishing its SUMOylation. The inhibition was shown to be dependent on both BGLF4 SUMO binding and BGLF4 kinase activity. Global SUMOylation was suppressed by active BGLF4, but not by SIM or kinase-dead BGLF4 mutants. Furthermore, interaction of BGLF4 with SUMO was required to induce DDR and enhance extracellular viral production during EBV lytic replication.

Identification of novel streptococcal proteins that bind human ligands

Pathogen–host interactions are not just limited to viruses, and the need to find novel antibiotics is becoming increasingly pertinent. Therefore, it is not surprising that protein microarrays have been utilized to identify bacterial pathogen–host interactions. Margarit and colleagues used microarrays to identify proteins expressed by two species of the streptococcus Gram-positive bacteria, *Streptococcus pyogenes* and *S. agalactiae,* that interact with human factors known to mediate pathogenesis (Margarit *et al.*, 2009). A bioinformatics approach was first used to predict 200 proteins that are present on the cellular surface, and therefore are more likely to paly a role in infection. The human probes were also carefully selected because of their known roles in colonization and infection: fibronectin, fibrinogen, and C4-binding protein. Treating the array consisting of the 200 predicted proteins with these probes identified 17 of the 20 known interactions, as well as 8 novel contacts, which were further validated using far-Western blot analysis. Three of the novel proteins identified were related, termed fib proteins, and using domain mapping, the authors were able to identify regions of these proteins that were required for their interaction with the human ligands. Subsequent assays of sera from patients with *S. agalactiae* showed high titres of Fib-specific antibodies, indicating their relative abundance during infection. There is hope that future studies will be able to determine the exact role of these proteins in infection and if they are suitable drug targets for *Streptococcus* infections.

Taken together, these studies show the tremendous power of protein microarrays in the discovery of novel molecular mechanisms underlying host–pathogen interactions.

Cancer research

Over the last 5 years, rapid development of genome-wide sequencing (e.g. next-generation sequencing) has revealed the heterogeneous nature of tumours, but clinical diagnosis is largely still dependent on morphological patterns (Parsons *et al.*, 2011; Gerlinger *et al.*, 2012). Although tumours have indistinguishable morphology, they often have vastly different clinical outcomes, so the composition of this heterogeneity needs to be better understood before more effective therapies can be developed. An individualistic therapy approach is thought to be the future of cancer research, so a new class of proteomic profiling technologies will have to be developed in order to meet this need. Protein microarrays have shown the potential to do just that and have proven useful in profiling the functional state of tumours and cancer biomarker identification (Table 5.6).

Reverse phase protein microarrays in cancer biology

One common approach that has been used to determine the status of signalling pathways in tumour cells is based on an immunoblot assay that uses antibodies to recognize phosphorylated proteins. Haab *et al.* (2001) was the first to scale this up to high throughput when they developed an antibody array, spotting individual commercial antibodies onto glass slides

Table 5.6 Summary of cancer studies using protein microarrays

Cancer type	Substrate	Probe type	Findings	Reference
Bladder	8000 human proteins	Patient sera	171 serum autoantibodies	Orenes-Pinero *et al.* (2010)
	254 serum protein antibodies	Patient sera	Protein profiles (including C-Met)	Sanchez-Carbayo *et al.* (2006)
Breast	4988 candidate tumour antigens	Patient sera	119 serum autoantibodies	Anderson *et al.* (2011)
	94 lectins	Mammalian cell lines	Cancer stem-cell like glycan signatures	Tao *et al.* (2008)
	Cytokine antibody array	MGF7/Her2–18 cell line	Cytokine signatures from breast cancer cell line	Vazquez-Martin *et al.* (2007)
	378 antibody array	Malignant breast tissue	Cancer expression profiles	Hudelist *et al.* (2004)
Lung	378 antibody array	Patient tissue samples	29 differentially regulated targets	Bartling *et al.* (2005)
Pancreatic	48 antibody array	Patient sera	Glycan-alterations of MUC1 and CEA	Chen *et al.* (2007)
Prostate	184 antibody array	Patient sera	Five differentially regulated targets	Miller *et al.* (2003)
	107 antibody array	Patient sera	Several targets, including Trombospondin-1	Shafer *et al.* (2007)
Rhabdomyo-sarcoma	59 cancer cell lysates	Phospho-specific antibodies	Phosphorylation status of 27 proteins	Petricoin *et al.* (2007)

in high density. This technology allows for the simultaneous detection of multiple antigens presented in a complex biological sample, such as cells, tissues, and body fluids (Borrebaeck *et al.*, 2009; Haab, 2003).

In 2001, Paweletz and colleagues coined the term 'reverse phase protein microarray' to describe an array in which lysates of cells or tissues are immobilized on the array surface rather than antibodies (Paweletz *et al.*, 2001). By using phosphoprotein-specific antibodies, these arrays can be used to interrogate the phosphorylation state present in cell lysates from tumour samples. Many clinical trials are currently ongoing that are utilizing these arrays (Muellet *et al.*, 2010). A large hurdle for most clinical trials and diagnostic tests is the minimal amount of sample that is available from the patients. Unobtrusive methods are few and far between, so these arrays are a welcome approach because not only can it allow for the evaluation of multiple components of a signalling pathway, very little sample is required for microarray fabrication. These samples can be printed on a series of identical arrays and analysed in parallel using commercially available phosphoprotein-specific antibodies.

Petricoin and coworkers obtained 59 cell lysate samples from the Children's Oncology Group Intergroup Rhabdomyosarcoma Study (IRS), and fabricated them into a reverse phase microarray (Petricoin *et al.*, 2007). Rhabdomyosarcoma is a rare childhood cancer that arises from undifferentiated muscle progenitor cells. Current treatment can yield a disease-free survival rate of close to 67%, but the reason for the failure of the treatment for the remaining one-third of patients is largely misunderstood. Identification of biomarkers could help distinguish these patients from the responders to the traditional therapy, which would aid in pursuing better therapies, and possibly identify other novel potential drug targets. The reverse phase microarrays were used as a platform to detect the phosphorylation status of proteins (using phospho-specific antibodies) thought to be distinguishing factors between the rhabdomyosarcoma subtypes. The group was able to identify higher phosphorylation levels in 4 Akt/mTOR pathway components that were only in patients with poor survival outcomes. Conversely, through bioinformatics analysis, they found that patients with good treatment outcomes exhibited mTOR pathway suppression. Together, these findings suggest that pharmacologically suppressing the mTOR pathway could result in improved outcomes for patients who did not respond to the standard chemotherapy treatment. Using a mouse xenograft model of rhabdomyosarcoma cell lines, there were able to test this hypothesis with known mTOR pathway inhibitors. As expected, treatment with the inhibitor resulted in reduced phosphorylation of the protein 4E-BP1, a protein that had been identified in the array screen, as well as reduced tumour growth. As a further proof of principle, they were able to demonstrate with the cell lysate arrays that specific patient sub-populations could be identified that benefited from personalized therapies and were able to identify the specific molecules that should be targeted in order to test these therapies.

Although this technique holds promise, there are several major problems with this approach. First, well characterized antibodies are not available for a great majority of human proteins, especially phosphoprotein-specific ones (Kalyuzhny, 2009; Couchman, 2009). Second, recent studies have advocated that many of the commercially available monoclonal antibodies may not even recognize their claimed targets, and can cross-react extensively with other cellular components (Jensen *et al.*, 2009). Third, when using antibodies for diagnostic and therapeutic applications, cross-reactivity is a large concern, as indicated by the recent withdrawal of several mAb-based pharmaceuticals from the market (Hughes, 2010; Berger *et al.*, 2009). Finally, this technique is not ideal for the discovery of novel biomarkers, as prior knowledge and phospho-specific antibodies are required for any evaluation.

Identification of autoantibody biomarkers for the early detection of breast cancer

While breast cancer awareness has resulted in many saved lives, current screening using mammograms detects only 70% of breast cancers, and false-positive mammograms leads to painful and unnecessary biopsies. If one were to identify biomarkers allowing for early detection of breast cancer, it could provide a non-invasive and low cost method that also improves patient outcomes. One arising category of cancer biomarkers is autoantibodies to tumour antigens. They offer better stability, specificity, ease of purification and detection compared to other serum proteins. Anderson *et al.* (2011) treated NAPPA protein arrays (described previously) containing tumour antigens with either breast cancer patient or control serum samples, in order to identify any differences in the human antibody repertoire that could be used as a biomarker. The first step in identifying new biomarkers used arrays containing 4988 tumour antigens in order to eliminate uninformative autoantibodies that were present in both patient and control samples. After identifying these background antigens, the number was reduced to 761, which allowed them to fabricate much smaller arrays for the next phase of study that were cheaper and would contain less false positives. In this next phase, sera from patients with invasive early breast cancer and benign breast disease were compared, resulting in 119 antigens that were present only in invasive cancer. The third phase validated 28 antigens that maintained high levels of specificity in a blinded assay, one of which was a known autoantigen ATP6AP1. They were able to show high expression of ATP6AP1 in four breast cancer cell lines by western blot as well as higher ATP6AP1 autoantibody levels in approximately 13% of early breast cancer serum samples compared to control. While only a first step, this work demonstrates the usefulness of protein microarrays, particularly NAPPA arrays, in the identification of early biomarkers in breast cancer.

Finding autoantibody biomarkers in bladder cancer

The utility of cancer biomarkers is in the development of new strategies for early diagnosis of disease, which allows for early intervention with current therapies and the improvement of patient survival rates. Cancer associated autoantibodies often target proteins that are mutated, modified, or aberrantly expressed in tumour cells, so they can be useful immunologic reporters that identify the molecular mechanisms underlying tumorigenesis. This, in turn, may represent a good starting point for the development of new treatments. With regards to bladder cancer, Orenes-Piñero and colleagues sought to identify not only autoantibody biomarkers of early bladder cancer, but also understand the underlying pathology of the disease (Orenes-Piñero *et al.*, 2010). Comparing serum samples collected from 12 patients with bladder cancer with 10 control patients using the Invitrogen Protoarray (an array consisting of ≈ 8000 purified human proteins), they identified 171 differentially expressed proteins.

Two proteins, clusterin and dynamin, were among the identified and were further validated using a custom tissue microarray comprised of bladder cancer tumour samples. They applied immunohistochemistry to these arrays and found reduced expression levels of clusterin in muscle invasive bladder cancer, compared to non-muscle invasive tumours. It was also shown that low expression of dynamin was associated with increased tumour stage and grade, higher recurrence rate after surgery, and shorter survival. Subsequent findings associating low levels of both clusterin and dynamin with disease appeared contradictory to the initial observation, which showed an increase in autoantibody levels to these proteins in patients with bladder cancer. However, despite these contrary findings, they were able to demonstrate significant associations between clusterin and dynamin protein levels and disease progression, and still potentially make use of them as biomarkers in the clinic and for drug screens.

Data analysis and bioinformatics

One common characteristic of all microarray experiments is the generation of large quantities of data. In order to elucidate interactions, careful statistical analysis is often required. This is most commonly performed by bioinformaticians, who work in close collaboration with laboratory scientists to make sense of the vast pools of data generated from a typical experiment. The analysis of data generated from protein microarray studies can be divided into two stages, or levels. First level analysis involves obtaining a reliable signal intensity value for each spot on a protein microarray for a particular assay. Second-level analysis is more advanced analysis built upon the profile of signal intensity values obtained from the first-level analysis.

First level analysis

The analysis at this level includes data acquisition, data normalization, and identification of hits. The raw data directly acquired from protein microarrays contain noise, which arise from various sources, such as inconsistent printing, uneven distribution of the reaction mixtures, inconsistent washing conditions, and batch effects. In most cases, algorithms developed for DNA/oligo microarray analysis can be applied to protein microarray analysis (Smyth *et al.*, 2003). It is often the case that proteins in a local region of the microarray show higher signal intensity than the overall intensity of the entire array. Since proteins are usually printed in a random order on a protein microarray, without regard for biological structure or function, it is not expected that such an effect is biological. To correct for this noise, the most useful normalization is the local normalization algorithm (e.g. ProCAT; Zhu *et al.*, 2006). For each spot, the average intensity of a small set of proteins (e.g. a 3-by-3 grid with the spot of interest in the middle) surrounding the spot as the local background intensity is used. Next, the local background is subtracted from the intensity of the spot of interest to achieve local noise correction. This process is then repeated for all spots on a protein microarray.

The next step is to identify positive hits for a particular assay. Because of the high cost associated with protein microarrays, replicates are often not employed, so a vigorous significance (e.g. *P*-value) for each protein can generally not be calculated. Rather, the overall intensity distribution of proteins on a protein microarray is calculated. In many cases, the distribution can be fit by a normal distribution curve. Proteins with intensity above a certain number of standard deviations (e.g. 3 SDs) are considered positive hits. In some situations, a set of negative control experiments can be performed, generating intensity profiles that can be used to estimate the false discovery rate.

Second level analysis

Various bioinformatics tools are utilized or developed for the second level analyses. Using biomarker identification as an example, the main question asked involves how to classify the groups of clustered proteins in order to draw conclusions. In other words, the computational problem is identifying an optimal set of proteins that can best separate the disease from the control samples. Normally, if one can identify elevated autoantibodies in disease samples relative to the controls, the corresponding autoantigens (i.e. human proteins) can serve as biomarkers for the disease.

There are two major steps of the classification: (1) selection of a set of proteins, and (2) evaluation of the performance of this combination of these proteins. Simply using the most antigenic protein may not be meaningful because its specificity and sensitivity is often not sufficient for diagnosis. In addition, a brutal force search for protein combinations is also not feasible. Different machine learning algorithms can be used to select the best combinations. Examples of such models include those based on multivariate logistic regression, mixtures of Gaussian distributions (Skates *et al.*, 2004), decision trees, artificial neural networks (ANNs), support vector machines (SVMs), and various approaches to incorporate these models in some kind of ensembles to improve overall performance (Zhang *et al.*, 2007, 2010). In addition, heuristic searching algorithms, such as sequential forward search and sequential forward floating search, can also be used to search for the best combination of features (i.e. proteins) (Sahiner *et al.*, 2000).

The performance will be assessed by sensitivity, specificity, and receiver-operating characteristic (ROC). To derive models that are robust and hence, more likely biologically based, and to alleviate the constraint of typical suboptimal sample sizes, statistical re-sampling methods, such as bootstrap, k-fold or leave-one-out cross-validation and introduction of artificial perturbations, have been widely utilized. These combined with known databases and gene ontology (GO) analysis, narrow down hits to more biologically relevant targets that can be further characterized in the laboratory.

Outlook and perspective

The development of the protein microarray over the past decade has led to large strides in our scientific knowledge. Combining the versatility and specificity with the high-throughput nature of the technology allows many labs to study anything from protein binding interactions, to post-translational modifications, to biomarkers quickly and reproducibly.

While some studies have identified protein modifications using MS/MS approaches and even protein microarrays (Newman *et al.*, 2013), there is still a great deal left unknown about how an enzyme specifically targets its substrate. Peptide arrays are one growing field with a promising outlook on motif identification. Studies by Rothbart and Mah have utilized peptide arrays to identify histone modifications and PTM motifs (Rothbart *et al.*, 2012; Mah *et al.*, 2005). The fabrication techniques still leave room for improvement, however, as using *in vitro* translated peptides is expensive and slow, and *in situ* array formation lacks in quality control. *In situ* arrays have the most straightforward approach to improvement, but this subset of microarrays will still be a powerful tool for modification identification in the future.

Although useful, to date many studies have used microarrays with limited proteomic coverage, limiting the characterization of what is truly a 'proteome' array. This is especially true in higher eukaryotes. The Zhu laboratory has taken great strides in increasing this coverage with their human proteome array, which consists of over 17,000 full-length human proteins, and is the largest to date (Jeong *et al.*, 2012). Utilizing this array could lead to further understanding and deeper characterization of the many PTM that have been identified via MS/MS approaches. It is believed that these technologies will widely be used in the identification of biomarkers as well. Recent studies have focused more on infectious and autoimmune diseases, but can be greatly useful in the identification of cancer biomarkers, potentially leading to diagnostic and prognostic capabilities. The human proteome array is the ideal array to use for this technique, as it currently is the most comprehensive way to screen for antigens. Although large screens can be prohibitively expensive, using the two-phase strategy in the AIH study described earlier (Song *et al.*, 2010) could be the approach used clinically to obtain the greatest results. Using mammalian cell lysates on these arrays could also provide a more relevant pseudo-*in vivo* system, building upon the technique utilized by Merbl and Kirschner (2009). The benefit of having reaction cofactors in the reaction mix cannot be overlooked, which should drastically reduce the number of false positives and false negatives that can plague typical studies.

Perhaps the most exciting new application in the study of microarrays is the advent of SPR and OI-RD technologies. While still in its infancy, many experiments can be imagined that utilize this technique. One of the largest hurdles in using antibody arrays and in the identification of biomarkers is the dearth of specific commercial antibodies. To date, there is no standard metric for antibody affinities used for commercially available antibodies, which always results in a series of optimization steps by each lab in order to determine the best conditions for a particular assay. Using SPR/OI-RD approaches could easily be used to screen antibody affinities and, in turn, create some sort of unified measurement of affinity. These label-free approaches could also theoretically be used for large-scale drug screens, without compromising the binding efficiency through conjugation with fluorophores or other large molecules such as biotin. One of the largest benefits of these approaches, however, is the ability to measure these interactions in real time. That, coupled with the high throughput nature of microarrays, can provide a powerful system of measuring enzyme kinetics and dynamics, drastically increasing our knowledge of these interactions.

While much work is still to be done in truly understanding our world, protein microarrays appear up to the task for large-scale proteomic studies. No other system allows for this amount of sensitivity, versatility, and high throughput ability than a microarray. It is our belief that the future of personalized medicine will also utilize this technology in one form or another, leading to better diagnosis and prognosis in many diseases. As this technology continues to mature, we can look forward to all the new connections that will be made, for protein microarrays are truly a versatile tool for scientific discovery.

References

Anderson, K.S., Sibani, S., Wallstrom, G., Qiu, J., Mendoza, E.A., Raphael, J., Hainsworth, E., Montor, W.R., Wong, J., Park, J.G., *et al.* (2011). Protein microarray signature of autoantibody biomarkers for the early detection of breast cancer. J. Proteome Res. *10*, 85–96.

Angenendt, P., Kreutzberger, J., Glokler, J., and Hoheisel, J.D. (2006). Generation of high density protein microarrays by cell-free *in situ* expression of unpurified PCR products. Mol. Cell. Proteomics *5*, 1658–1666.

Barbulovic-Nad, I., Lucente, M., Sun, Y., Zhang, M., Wheeler, A.R., and Bussmann, M. (2006). Biomicroarray fabrication techniques – a review. Crit. Rev. Biotechnol. *26*, 237–259.

Bartling, B., Hofmann, H.S., Boettger, T., Hansen, G., Burdach, S., Silber, R.E., and Simm, A. (2005). Comparative application of antibody and gene array for expression profiling in human squamous cell lung carcinoma. Lung Cancer *49*, 145–154.

Berger, J.R., Houff, S.A., and Major, E.O. (2009). Monoclonal antibodies and progressive multifocal leukoencephalopathy. Mabs *1*, 583–589.

Borrebaeck, C.A., and Wingren, C. (2009). Design of high-density antibody microarrays for disease proteomics: key technological issues. J. Proteomics *72*, 928–935.

Brichta, J., Hnilova, M., and and Viskovic, T. (2005). Generation of hapten-specific recombinant antibodies: Antibody phage display technology: A review. Vet. Med. *50*, 231–252.

Bussow, K., Cahill, D., Nietfeld, W., Bancroft, D., Scherzinger, E., Lehrach, H., and Walter, G. (1998). A method for global protein expression and antibody screening on high-density filters of an arrayed cDNA library. Nucleic Acids Res. *26*, 5007–5008.

Carmen, S., and Jermutus, L. (2002). Concepts in antibody phage display. Brief Funct. Genomic Proteomic *1*, 189–203.

Chang, T.W. (1983). Binding of cells to matrixes of distinct antibodies coated on solid surface. J. Immunol. Methods *65*, 217–223.

Chao, G., Lau, W.L., Hackel, B.J., Sazinsky, S.L., Lippow, S.M., and Wittrup, K.D. (2006). Isolating and engineering human antibodies using yeast surface display. Nat. Protoc. *1*, 755–768.

Chen, C.S., and Zhu, H. (2006). Protein microarrays. Biotechniques *40*, 423, 425, 427 passim.

Chen, C.S., Korobkova, E., Chen, H., Zhu, J., Jian, X., Tao, S.C., He, C., and Zhu, H. (2008). A proteome chip approach reveals new DNA damage recognition activities in Escherichia coli. Nat. Methods *5*, 69–74.

Chen, C.S., Sullivan, S., Anderson, T., Tan, A.C., Alex, P.J., Brant, S.R., Cuffari, C., Bayless, T.M., Talor, M.V., Burek, C.L., *et al.* (2009). Identification of novel serological biomarkers for inflammatory bowel disease using Escherichia coli proteome chip. Mol. Cell. Proteomics *8*, 1765–1776.

Chen, S., LaRoche, T., Hamelinck, D., Bergsma, D., Brenner, D., Simeone, D., Brand, R.E., and Haab, B.B. (2007). Multiplexed analysis of glycan variation on native proteins captured by antibody microarrays. Nat. Methods *4*, 437–444.

Couchman, J.R. (2009). Commercial antibodies: the good, bad, and really ugly. J. Histochem. Cytochem. *57*, 7–8.
Diamond, D.L., Zhang, Y., Gaiger, A., Smithgall, M., Vedvick, T.S., and Carter, D. (2003). Use of ProteinChip array surface enhanced laser desorption/ionization time-of-flight mass spectrometry (SELDI-TOF MS) to identify thymosin beta-4, a differentially secreted protein from lymphoblastoid cell lines. J. Am. Soc. Mass Spectrom. *14*, 760–765.
Dias, W.B., Cheung, W.D., and Hart, G.W. (2012). O-GlcNAcylation of kinases. Biochem. Biophys. Res. Commun. *422*, 224–228.
Doolan, D.L., Mu, Y., Unal, B., Sundaresh, S., Hirst, S., Valdez, C., Randall, A., Molina, D., Liang, X., Freilich, D.A., *et al.* (2008). Profiling humoral immune responses to P. falciparum infection with protein microarrays. Proteomics *8*, 4680–4694.
Ekins, R.P. (1989). Multi-analyte immunoassay. J. Pharm. Biomed. Anal. *7*, 155–168.
Evans-Nguyen, K.M., Tao, S.C., Zhu, H., and Cotter, R.J. (2008). Protein arrays on patterned porous gold substrates interrogated with mass spectrometry: detection of peptides in plasma. Anal. Chem. *80*, 1448–1458.
Foster, M.W., Forrester, M.T., and Stamler, J.S. (2009). A protein microarray-based analysis of S-nitrosylation. Proc. Natl. Acad. Sci. U. S. A. *106*, 18948–18953.
Fu, D., Ma, J., and Chen, J. (2007). Dual-wavelength microarray fluorescence detection system using volume holographic filter. J. Biomed. Opt. *12*, 014040.
Gavin, I.M., Kukhtin, A., Glesne, D., Schabacker, D., and Chandler, D.P. (2005). Analysis of protein interaction and function with a 3-dimensional MALDI-MS protein array. Biotechniques *39*, 99–107.
Ge, H. (2000). UPA, a universal protein array system for quantitative detection of protein-protein, protein-DNA, protein-RNA and protein–ligand interactions. Nucleic Acids Res. *28*, e3.
Gupta, R., Kus, B., Fladd, C., Wasmuth, J., Tonikian, R., Sidhu, S., Krogan, N.J., Parkinson, J., and Rotin, D. (2007). Ubiquitination screen using protein microarrays for comprehensive identification of Rsp5 substrates in yeast. Mol. Syst. Biol. *3*, 116.
Gutmann, O., Kuehlewein, R., Reinbold, S., Niekrawietz, R., Steinert, C.P., de Heij, B., Zengerle, R., and Daub, M. (2004). A highly parallel nanoliter dispenser for microarray fabrication. Biomed. Microdevices *6*, 131–137.
Guttman, M., Amit, I., Garber, M., French, C., Lin, M.F., Feldser, D., Huarte, M., Zuk, O., Carey, B.W., Cassady, J.P., *et al.* (2009). Chromatin signature reveals over a thousand highly conserved large non-coding RNAs in mammals. Nature *458*, 223–227.
Gygi, S.P., Rochon, Y., Franza, B.R., and Aebersold, R. (1999). Correlation between protein and mRNA abundance in yeast. Mol. Cell. Biol. *19*, 1720–1730.
Haab, B.B. (2003). Methods and applications of antibody microarrays in cancer research. Proteomics 3, 2116–2122.
Haab, B.B., Dunham, M.J., and Brown, P.O. (2001). Protein microarrays for highly parallel detection and quantitation of specific proteins and antibodies in complex solutions. Genome Biol. *2*, RESEARCH0004.
Hall, D.A., Zhu, H., Zhu, X., Royce, T., Gerstein, M., and Snyder, M. (2004). Regulation of gene expression by a metabolic enzyme. Science *306*, 482–484.
He, M., and Taussig, M.J. (2001). Single step generation of protein arrays from DNA by cell-free expression and *in situ* immobilisation (PISA method). Nucleic Acids Res. *29*, E73–3.
Ho, S.W., Jona, G., Chen, C.T., Johnston, M., and Snyder, M. (2006). Linking DNA-binding proteins to their recognition sequences by using protein microarrays. Proc. Natl. Acad. Sci. U. S. A. *103*, 9940–9945.
Hu, C.J., Song, G., Huang, W., Liu, G.Z., Deng, C.W., Zeng, H.P., Wang, L., Zhang, F.C., Zhang, X., Jeong, J.S., *et al.* (2012). Identification of new autoantigens for primary biliary cirrhosis using human proteome microarrays. Mol. Cell. Proteomics *11*, 669–680.
Hu, J., Rho, H.S., Newman, R.H., Hwang, W., Neiswinger, J., Zhu, H., Zhang, J., and Qian, J. (2013). Global analysis of phosphorylation networks in humans. Biochim. Biophys. Acta
Hu, S., Li, Y., Liu, G., Song, Q., Wang, L., Han, Y., Zhang, Y., Song, Y., Yao, X., Tao, Y., *et al.* (2007). A protein chip approach for high-throughput antigen identification and characterization. Proteomics *7*, 2151–2161.
Hu, S., Xie, Z., Onishi, A., Yu, X., Jiang, L., Lin, J., Rho, H.S., Woodard, C., Wang, H., Jeong, J.S., *et al.* (2009). Profiling the human protein-DNA interactome reveals ERK2 as a transcriptional repressor of interferon signaling. Cell *139*, 610–622.
Hu, S., Vissink, A., Arellano, M., Roozendaal, C., Zhou, H., Kallenberg, C.G., and Wong, D.T. (2011). Identification of autoantibody biomarkers for primary Sjogren's syndrome using protein microarrays. Proteomics *11*, 1499–1507.

Hu, S., Xie, Z., Qian, J., Blackshaw, S., and Zhu, H. (2011). Functional protein microarray technology. Wiley Interdiscip. Rev. Syst. Biol. Med. *3*, 255–268.

Huang, J., Zhu, H., Haggarty, S.J., Spring, D.R., Hwang, H., Jin, F., Snyder, M., and Schreiber, S.L. (2004). Finding new components of the target of rapamycin (TOR) signaling network through chemical genetics and proteome chips. Proc. Natl. Acad. Sci. U S.A. *101*, 16594–16599.

Hudelist, G., Pacher-Zavisin, M., Singer, C.F., Holper, T., Kubista, E., Schreiber, M., Manavi, M., Bilban, M., and Czerwenka, K. (2004). Use of high-throughput protein array for profiling of differentially expressed proteins in normal and malignant breast tissue. Breast Cancer Res. Treat. *86*, 281–291.

Hughes, B. (2010). Antibody–drug conjugates for cancer: poised to deliver? Nat. Rev. Drug Discov. *9*, 665–667.

Jensen, B.C., Swigart, P.M., and Simpson, P.C. (2009). Ten commercial antibodies for alpha-1-adrenergic receptor subtypes are nonspecific. Naunyn Schmiedebergs Arch. Pharmacol. *379*, 409–412.

Jeong, J.S., Jiang, L., Albino, E., Marrero, J., Rho, H.S., Hu, J., Hu, S., Vera, C., Bayron-Poueymiroy, D., Rivera-Pacheco, Z.A., *et al.* (2012). Rapid identification of monospecific monoclonal antibodies using a human proteome microarray. Mol. Cell. Proteomics *11*, O111.016253.

Jones, R.B., Gordus, A., Krall, J.A., and MacBeath, G. (2006). A quantitative protein interaction network for the ErbB receptors using protein microarrays. Nature *439*, 168–174.

Joung, J.K., Ramm, E.I., and Pabo, C.O. (2000). A bacterial two-hybrid selection system for studying protein-DNA and protein–protein interactions. Proc. Natl. Acad. Sci. U.S.A. *97*, 7382–7387.

Kalyuzhny, A.E. (2009). The dark side of the immunohistochemical moon: industry. J. Histochem. Cytochem. *57*, 1099–1101.

Kodoyianni, V. (2011). Label-free analysis of biomolecular interactions using SPR imaging. Biotechniques *50*, 32–40.

Kopf, E., and Zharhary, D. (2007). Antibody arrays – an emerging tool in cancer proteomics. Int. J. Biochem. Cell Biol. *39*, 1305–1317.

Krogan, N.J., Cagney, G., Yu, H., Zhong, G., Guo, X., Ignatchenko, A., Li, J., Pu, S., Datta, N., Tikuisis, A.P., *et al.* (2006). Global landscape of protein complexes in the yeast Saccharomyces cerevisiae. Nature *440*, 637–643.

Kung, L.A., Tao, S.C., Qian, J., Smith, M.G., Snyder, M., and Zhu, H. (2009). Global analysis of the glycoproteome in Saccharomyces cerevisiae reveals new roles for protein glycosylation in eukaryotes. Mol. Syst. Biol. *5*, 308.

Kuno, A., Kato, Y., Matsuda, A., Kaneko, M.K., Ito, H., Amano, K., Chiba, Y., Narimatsu, H., and Hirabayashi, J. (2009). Focused differential glycan analysis with the platform antibody-assisted lectin profiling for glycan-related biomarker verification. Mol. Cell. Proteomics *8*, 99–108.

Landry, J.P., Sun, Y.S., Guo, X.W., and Zhu, X.D. (2008). Protein reactions with surface-bound molecular targets detected by oblique-incidence reflectivity difference microscopes. Appl. Opt. *47*, 3275–3288.

Li, B., Jiang, L., Song, Q., Yang, J., Chen, Z., Guo, Z., Zhou, D., Du, Z., Song, Y., Wang, J., *et al.* (2005). Protein microarray for profiling antibody responses to Yersinia pestis live vaccine. Infect. Immun. *73*, 3734–3739.

Li, R., Zhu, J., Xie, Z., Liao, G., Liu, J., Chen, M.R., Hu, S., Woodard, C., Lin, J., Taverna, S.D., *et al.* (2011). Conserved herpesvirus kinases target the DNA damage response pathway and TIP60 histone acetyltransferase to promote virus replication. Cell. Host Microbe *10*, 390–400.

Li, R., Wang, L., Liao, G., Guzzo, C.M., Matunis, M.J., Zhu, H., and Hayward, S.D. (2012). SUMO binding by the Epstein–Barr virus protein kinase BGLF4 is crucial for BGLF4 function. J. Virol. *86*, 5412–5421.

Li, Y., Tao, S.C., Bova, G.S., Liu, A.Y., Chan, D.W., Zhu, H., and Zhang, H. (2011). Detection and verification of glycosylation patterns of glycoproteins from clinical specimens using lectin microarrays and lectin-based immunosorbent assays. Anal. Chem. *83*, 8509–8516.

Liang, L., Tan, X., Juarez, S., Villaverde, H., Pablo, J., Nakajima-Sasaki, R., Gotuzzo, E., Saito, M., Hermanson, G., Molina, D., *et al.* (2011). Systems biology approach predicts antibody signature associated with *Brucella melitensis* infection in humans. J. Proteome Res. *10*, 4813–4824.

Lin, Y.Y., Lu, J.Y., Zhang, J., Walter, W., Dang, W., Wan, J., Tao, S.C., Qian, J., Zhao, Y., Boeke, J.D., Berger, S.L., and Zhu, H. (2009). Protein acetylation microarray reveals that NuA4 controls key metabolic target regulating gluconeogenesis. Cell *136*, 1073–1084.

Lu, J.Y., Lin, Y.Y., Qian, J., Tao, S.C., Zhu, J., Pickart, C., and Zhu, H. (2008). Functional dissection of a HECT ubiquitin E3 ligase. Mol. Cell. Proteomics *7*, 35–45.

Lu, K.Y., Tao, S.C., Yang, T.C., Ho, Y.H., Lee, C.H., Lin, C.C., Juan, H.F., Huang, H.C., Yang, C.Y., Chen, M.S., *et al.* (2012). Profiling lipid–protein interactions using nonquenched fluorescent liposomal nanovesicles and proteome microarrays. Mol. Cell. Proteomics *11*, 1177–1190.

Luevano, M., Bernard, H.U., Barrera-Saldana, H.A., Trevino, V., Garcia-Carranca, A., Villa, L.L., Monk, B.J., Tan, X., Davies, D.H., Felgner, P.L., and Kalantari, M. (2010). High-throughput profiling of the humoral immune responses against thirteen human papillomavirus types by proteome microarrays. Virology *405*, 31–40.

MacBeath, G., and Schreiber, S.L. (2000). Printing proteins as microarrays for high-throughput function determination. Science *289*, 1760–1763.

Mah, A.S., Elia, A.E., Devgan, G., Ptacek, J., Schutkowski, M., Snyder, M., Yaffe, M.B., and Deshaies, R.J. (2005). Substrate specificity analysis of protein kinase complex Dbf2-Mob1 by peptide library and proteome array screening. BMC Biochem. *6*, 22.

Margarit, I., Bonacci, S., Pietrocola, G., Rindi, S., Ghezzo, C., Bombaci, M., Nardi-Dei, V., Grifantini, R., Speziale, P., and Grandi, G. (2009). Capturing host–pathogen interactions by protein microarrays: identification of novel streptococcal proteins binding to human fibronectin, fibrinogen, and C4BP. Faseb j. *23*, 3100–3112.

Merbl, Y., and Kirschner, M.W. (2009). Large-scale detection of ubiquitination substrates using cell extracts and protein microarrays. Proc. Natl. Acad. Sci. U. S. A. *106*, 2543–2548.

Miller, J.C., Zhou, H., Kwekel, J., Cavallo, R., Burke, J., Butler, E.B., Teh, B.S., and Haab, B.B. (2003). Antibody microarray profiling of human prostate cancer sera: antibody screening and identification of potential biomarkers. Proteomics *3*, 56–63.

Newman, R.H., Hu, J., Rho, H.S., Xie, Z., Woodard, C., Neiswinger, J., Cooper, C., Shirley, M., Clark, H.M., Hu, S., *et al.* (2013). Construction of human activity-based phosphorylation networks. Mol. Syst. Biol. *9*, 655.

Nikitin, P.A., and Luftig, M.A. (2012). The DNA damage response in viral-induced cellular transformation. Br. J. Cancer *106*, 429–435.

Oh, Y.H., Hong, M.Y., Jin, Z., Lee, T., Han, M.K., Park, S., and Kim, H.S. (2007). Chip-based analysis of SUMO (small ubiquitin-like modifier) conjugation to a target protein. Biosens. Bioelectron. *22*, 1260–1267.

Orenes-Pinero, E., Barderas, R., Rico, D., Casal, J.I., Gonzalez-Pisano, D., Navajo, J., Algaba, F., Piulats, J.M., and Sanchez-Carbayo, M. (2010). Serum and tissue profiling in bladder cancer combining protein and tissue arrays. J. Proteome Res. *9*, 164–173.

Parsons, D.W., Li, M., Zhang, X., Jones, S., Leary, R.J., Lin, J.C., Boca, S.M., Carter, H., Samayoa, J., Bettegowda, C., *et al.* (2011). The genetic landscape of the childhood cancer medulloblastoma. Science *331*, 435–439.

Paweletz, C.P., Charboneau, L., Bichsel, V.E., Simone, N.L., Chen, T., Gillespie, J.W., Emmert-Buck, M.R., Roth, M.J., Petricoin III, E.F., and Liotta, L.A. (2001). Reverse phase protein microarrays which capture disease progression show activation of pro-survival pathways at the cancer invasion front. Oncogene *20*, 1981–1989.

Petricoin, E.F.,3rd, Espina, V., Araujo, R.P., Midura, B., Yeung, C., Wan, X., Eichler, G.S., Johann, D.J.,Jr, Qualman, S., Tsokos, M., *et al.* (2007). Phosphoprotein pathway mapping: Akt/mammalian target of rapamycin activation is negatively associated with childhood rhabdomyosarcoma survival. Cancer Res. *67*, 3431–3440.

Poetz, O., Schwenk, J.M., Kramer, S., Stoll, D., Templin, M.F., and Joos, T.O. (2005). Protein microarrays: catching the proteome. Mech. Ageing Dev. *126*, 161–170.

Popescu, S.C., Popescu, G.V., Bachan, S., Zhang, Z., Seay, M., Gerstein, M., Snyder, M., and Dinesh-Kumar, S.P. (2007). Differential binding of calmodulin-related proteins to their targets revealed through high-density Arabidopsis protein microarrays. Proc. Natl. Acad. Sci. U. S. A. *104*, 4730–4735.

Popescu, S.C., Popescu, G.V., Bachan, S., Zhang, Z., Gerstein, M., Snyder, M., and Dinesh-Kumar, S.P. (2009). MAPK target networks in Arabidopsis thaliana revealed using functional protein microarrays. Genes Dev. *23*, 80–92.

Ptacek, J., Devgan, G., Michaud, G., Zhu, H., Zhu, X., Fasolo, J., Guo, H., Jona, G., Breitkreutz, A., Sopko, R., *et al.* (2005). Global analysis of protein phosphorylation in yeast. Nature *438*, 679–684.

Ramachandran, N., Hainsworth, E., Bhullar, B., Eisenstein, S., Rosen, B., Lau, A.Y., Walter, J.C., and LaBaer, J. (2004). Self-assembling protein microarrays. Science *305*, 86–90.

Rapicavoli, N.A., Poth, E.M., Zhu, H., and Blackshaw, S. (2011). The long noncoding RNA Six3OS acts in trans to regulate retinal development by modulating Six3 activity. Neural Dev. *6*, 32–8104–6–32.

Rothbart, S.B., Krajewski, K., Strahl, B.D., and Fuchs, S.M. (2012). Peptide microarrays to interrogate the 'histone code'. Methods Enzymol. *512*, 107–135.

Sahiner, B., Chan, H.P., Petrick, N., Wagner, R.F., and Hadjiiski, L. (2000). Feature selection and classifier performance in computer-aided diagnosis: the effect of finite sample size. Med. Phys. *27*, 1509–1522.

Salamat-Miller, N., Fang, J., Seidel, C.W., Smalter, A.M., Assenov, Y., Albrecht, M., and Middaugh, C.R. (2006). A network-based analysis of polyanion-binding proteins utilizing yeast protein arrays. Mol. Cell. Proteomics *5*, 2263–2278.

Sanchez-Carbayo, M., Socci, N.D., Lozano, J.J., Haab, B.B., and Cordon-Cardo, C. (2006). Profiling bladder cancer using targeted antibody arrays. Am. J. Pathol. *168*, 93–103.

Schnack, C., Hengerer, B., and Gillardon, F. (2008). Identification of novel substrates for Cdk5 and new targets for Cdk5 inhibitors using high-density protein microarrays. Proteomics *8*, 1980–1986.

Schweitzer, B., Roberts, S., Grimwade, B., Shao, W., Wang, M., Fu, Q., Shu, Q., Laroche, I., Zhou, Z., Tchernev, V.T., *et al.* (2002). Multiplexed protein profiling on microarrays by rolling-circle amplification. Nat. Biotechnol. *20*, 359–365.

Shafer, M.W., Mangold, L., Partin, A.W., and Haab, B.B. (2007). Antibody array profiling reveals serum TSP-1 as a marker to distinguish benign from malignant prostatic disease. Prostate *67*, 255–267.

Shamay, M., Liu, J., Li, R., Liao, G., Shen, L., Greenway, M., Hu, S., Zhu, J., Xie, Z., Ambinder, R.F., *et al.* (2012). A protein array screen for Kaposi's sarcoma-associated herpesvirus LANA interactors links LANA to TIP60, PP2A activity, and telomere shortening. J. Virol. *86*, 5179–5191.

Siprashvili, Z., Webster, D.E., Kretz, M., Johnston, D., Rinn, J.L., Chang, H.Y., and Khavari, P.A. (2012). Identification of proteins binding coding and non-coding human RNAs using protein microarrays. BMC Genomics *13*, 633–2164–13–633.

Skates, S.J., Horick, N., Yu, Y., Xu, F.J., Berchuck, A., Havrilesky, L.J., de Bruijn, H.W., van der Zee, A.G., Woolas, R.P., Jacobs, I.J., Zhang, Z., and Bast, R.C.,Jr. (2004). Preoperative sensitivity and specificity for early-stage ovarian cancer when combining cancer antigen CA-125II, CA 15–3, CA 72–4, and macrophage colony-stimulating factor using mixtures of multivariate normal distributions. J. Clin. Oncol. *22*, 4059–4066.

Smyth, G.K., and Speed, T. (2003). Normalization of cDNA microarray data. Methods *31*, 265–273.

Song, Q., Liu, G., Hu, S., Zhang, Y., Tao, Y., Han, Y., Zeng, H., Huang, W., Li, F., Chen, P., *et al.* (2010). Novel autoimmune hepatitis-specific autoantigens identified using protein microarray technology. J. Proteome Res. *9*, 30–39.

Stillman, B.A., and Tonkinson, J.L. (2000). FAST slides: a novel surface for microarrays. Biotechniques *29*, 630–635.

Tao, S.C., Li, Y., Zhou, J., Qian, J., Schnaar, R.L., Zhang, Y., Goldstein, I.J., Zhu, H., and Schneck, J.P. (2008). Lectin microarrays identify cell-specific and functionally significant cell surface glycan markers. Glycobiology *18*, 761–769.

Tarrant, M.K., Rho, H.S., Xie, Z., Jiang, Y.L., Gross, C., Culhane, J.C., Yan, G., Qian, J., Ichikawa, Y., Matsuoka, T., *et al.* (2012). Regulation of CK2 by phosphorylation and O-GlcNAcylation revealed by semisynthesis. Nat. Chem. Biol. *8*, 262–269.

Thao, S., Chen, C.S., Zhu, H., and Escalante-Semerena, J.C. (2010). Nepsilon-lysine acetylation of a bacterial transcription factor inhibits Its DNA-binding activity. PLoS One *5*, e15123.

Uchiyama, N., Kuno, A., Koseki-Kuno, S., Ebe, Y., Horio, K., Yamada, M., and Hirabayashi, J. (2006). Development of a lectin microarray based on an evanescent-field fluorescence principle. Methods Enzymol. *415*, 341–351.

Unfricht, D.W., Colpitts, S.L., Fernandez, S.M., and Lynes, M.A. (2005). Grating-coupled surface plasmon resonance: a cell and protein microarray platform. Proteomics *5*, 4432–4442.

Vazquez-Martin, A., Colomer, R., and Menendez, J.A. (2007). Protein array technology to detect HER2 (erbB-2)-induced 'cytokine signature' in breast cancer. Eur. J. Cancer *43*, 1117–1124.

Vidal, M., Brachmann, R.K., Fattaey, A., Harlow, E., and Boeke, J.D. (1996). Reverse two-hybrid and one-hybrid systems to detect dissociation of protein-protein and DNA–protein interactions. Proc. Natl. Acad. Sci. U. S. A. *93*, 10315–10320.

Vigil, A., Chen, C., Jain, A., Nakajima-Sasaki, R., Jasinskas, A., Pablo, J., Hendrix, L.R., Samuel, J.E., and Felgner, P.L. (2011). Profiling the humoral immune response of acute and chronic Q fever by protein microarray. Mol. Cell. Proteomics *10*, M110.006304.

Wegner, G.J., Lee, H.J., Marriott, G., and Corn, R.M. (2003). Fabrication of histidine-tagged fusion protein arrays for surface plasmon resonance imaging studies of protein-protein and protein–DNA interactions. Anal. Chem. *75*, 4740–4746.

Yan, H., Park, S.H., Finkelstein, G., Reif, J.H., and LaBean, T.H. (2003). DNA-templated self-assembly of protein arrays and highly conductive nanowires. Science *301*, 1882–1884.

Young, K.H. (1998). Yeast two-hybrid: so many interactions, (in) so little time.. Biol. Reprod. *58*, 302–311.

Zhang, X., Zhu, S., Deng, C., and Zhang, X. (2012). An aptamer based on-plate microarray for high-throughput insulin detection by MALDI-TOF MS. Chem. Commun. (Camb) *48*, 2689–2691.

Zhang, Z., Yu, Y., Xu, F., Berchuck, A., van Haaften-Day, C., Havrilesky, L.J., de Bruijn, H.W., van der Zee, A.G., Woolas, R.P., Jacobs, I.J., *et al.* (2007). Combining multiple serum tumor markers improves detection of stage I epithelial ovarian cancer. Gynecol. Oncol. *107*, 526–531.

Zhang, Z., and Chan, D.W. (2010). The road from discovery to clinical diagnostics: lessons learned from the first FDA-cleared *in vitro* diagnostic multivariate index assay of proteomic biomarkers. Cancer Epidemiol. Biomarkers Prev. *19*, 2995–2999.

Zheng, D., Wan, J., Cho, Y.G., Wang, L., Chiou, C.J., Pai, S., Woodard, C., Zhu, J., Liao, G., Martinez-Maza, O., *et al.* (2011). Comparison of humoral immune responses to Epstein–Barr virus and Kaposi's sarcoma-associated herpesvirus using a viral proteome microarray. J. Infect. Dis. *204*, 1683–1691.

Zhu, H., Klemic, J.F., Chang, S., Bertone, P., Casamayor, A., Klemic, K.G., Smith, D., Gerstein, M., Reed, M.A., and Snyder, M. (2000). Analysis of yeast protein kinases using protein chips. Nat. Genet. *26*, 283–289.

Zhu, H., Bilgin, M., Bangham, R., Hall, D., Casamayor, A., Bertone, P., Lan, N., Jansen, R., Bidlingmaier, S., Houfek, T., *et al.* (2001). Global analysis of protein activities using proteome chips. Science *293*, 2101–2105.

Zhu, H., Hu, S., Jona, G., Zhu, X., Kreiswirth, N., Willey, B.M., Mazzulli, T., Liu, G., Song, Q., Chen, P., *et al.* (2006). Severe acute respiratory syndrome diagnostics using a coronavirus protein microarray. Proc. Natl. Acad. Sci. U. S. A. *103*, 4011–4016.

Zhu, H., Cox, E., and Qian, J. (2012). Functional protein microarray as molecular decathlete: a versatile player in clinical proteomics. Proteomics Clin. Appl. *6*, 548–562.

Zhu, J., Gopinath, K., Murali, A., Yi, G., Hayward, S.D., Zhu, H., and Kao, C. (2007). RNA-binding proteins that inhibit RNA virus infection. Proc. Natl. Acad. Sci. U. S. A. *104*, 3129–3134.

Zhu, J., Liao, G., Shan, L., Zhang, J., Chen, M.R., Hayward, G.S., Hayward, S.D., Desai, P., and Zhu, H. (2009). Protein array identification of substrates of the Epstein–Barr virus protein kinase BGLF4. J. Virol. *83*, 5219–5231.

Zhu, X., Gerstein, M., and Snyder, M. (2006). ProCAT: a data analysis approach for protein microarrays. Genome Biol. *7*, R110.

Bioinformatics Challenges in Targeted Proteomics

6

Lars Malmström

Abstract

Targeted proteomics is becoming an important technology since it is capable of producing quantitative data with high reproducibility and accuracy for several proteins in a single experiment. The user needs to specify in great detail what to measure and the produced data are analysed using techniques that have been specially developed for this technology. Targeted proteomics is therefore fundamentally dependent on an extensive set of software tools at almost every step. This chapter describes the bioinformatics challenges in targeted proteomics and suggests possible tools which can be used in each step.

Introduction

Mass spectrometry-based proteomics is making important advances in areas such as instrumentation, software, experimental techniques and, importantly, acquisition strategies. These advances have resulted in faster data acquisition, greater flexibility and higher quality measurements. While the exact reason for collecting proteomics data varies from experiment to experiment, some hypotheses are more often addressed by mass spectrometry than others. The vast majority of these hypotheses are tested by determining differences between two or more samples. Differences in protein quantities and qualitative or quantitative differences in post-translational modifications are two of the established and heavily used technologies that have had a significant impact on several research fields. The challenge is to acquire data that allows for the extraction of the required information quickly and with high accuracy. The collected data need to be processed using one or more software tools to extract the desired information. Which tools to use varies significantly depending on the nature of the experiment and the fact that new tools are constantly being developed while, unfortunately, old tools disappear. This chapter is dedicated to the tools used to address the bioinformatics challenges in an acquisition technique called target proteomics. To keep the discussion focused and succinct, selected reaction monitoring (SRM) (Lange *et al.*, 2008) using triple-quadrupole instruments will be used in the examples throughout the chapter. However, the reader should be aware that several other targeted proteomics techniques and instruments are currently in use. The most common acquisition method by mass spectrometry is shotgun, which is briefly described in the paragraph below so that key differences with SRM can be highlighted.

Shotgun proteomics (Domon and Aebersold, 2006), data-dependent bottom-up acquisition, have long been the most commonly used acquisition strategy due to the relative ease of use and the ability to characterize the protein content of samples both in terms of protein identities and protein quantities (Weisser *et al.*, 2013). The most basic experimental set-up for shotgun proteomics can be summarized in a few steps: (i) samples of interest are digested using proteases such as trypsin; (ii) peptides are separated using high-pressure liquid chromatography and a reverse-phase column; (iii) the separated peptides are ionized using electro spray ionization (ESI); (vi) ionized peptides are surveyed by scanning the instrument to obtain intensities for a range of m/z values to generate an MS1 spectrum; (v) which peaks to fragment is determined based on the MS1 spectrum, along with which peaks have been previously measured; (vi) the selected peaks are fragmented and the instrument is scanned once more to produce a continuous measurement of the fragments producing an MS2 spectrum; (vii) peptides identified from the MS2 spectra are inferred using a search engine such as Sequest (Yates *et al.*, 1995), Mascot (Perkins *et al.*, 1999), OMSSA (Geer *et al.*, 2004) or X! tandem (Craig and Beavis, 2003); (viii) peptides are assembled and the results are statistically evaluated using peptideProphet (Keller *et al.*, 2002) and protein-Prophet (Nesvizhskii *et al.*, 2003) for example; and (ix) in quantitative studies, features are identified using the MS1 data and the peptide identities inferred from the MS2 are mapped onto the feature which, in the last step, is combined into protein quantities (Sturm *et al.*, 2008). Shotgun proteomics can be carried out on a wide range of instruments, but instruments that can rapidly scan wide mass ranges have an advantage (Olsen *et al.*, 2009).

The process for SRM is the same for steps (i)–(iii), but then differs significantly; the major differences are briefly mentioned here and then discussed in more detail throughout the rest of the chapter. By definition, targeted proteomics only acquires data for a set of predetermined peptides. The advantage is that this allows the instrument to spend most of the time acquiring the desired data. The obvious disadvantages include the fact that this does not acquire any data outside the set parameters, which in turn does not allow for any discovery. In SRM carried out on a triple-quadrupole, Q1 (step vi) is not scanned, but isolates a defined precursor ion which is then fragmented, step (v) is skipped, and Q3 (step vi) is not scanned either but is set to isolate a single fragment. As data are acquired for a particular peptide, there is no need to identify it; instead, the statistical operation aims estimate the signal from the peptide to as accurately as possible and remove the signal from noise that will invariably be present in the data. There is always the risk that the noise in the data can be mistaken for a signal from the peptide, which is often estimated as a false discovery rate and tries to describe how many of the signals in the dataset arise from something other than the peptide of interest. It is worth pointing out that step viii is not needed in SRM, as the quantity of the peptide is the main goal of the entire data acquisition and inferring peptide identities would be pointless. Most instruments are capable of generating targeted proteomics data, but some are more suited than others. For example, fast scanning is not necessarily an advantage. Instead, the ability to quickly switch between different isolation windows is advantageous and quadrupole-based instruments are therefore good choices for targeted proteomics in general and SRM in particular.

A few advantages and disadvantages have been mentioned so far, but it is worth making a more complete account. Shotgun or similar techniques will play a fundamental role in the foreseeable future as they accurately capture the lion's share of the most abundant peptide species resulting from the digestion of complex samples. No prior knowledge of the sample is necessary in order to capture the data, although knowing which organism is being investigated simplifies data analysis and thereby increases the sensitivity. Even modern instruments are not fast enough to identify all of the features present in the MS1 map, which can cause problems in downstream analysis due to missing quantities in a subset of samples. As most proteins produce many distinct peptide species and the mass spectrometer has no way of knowing which peptides came from the same protein, highly abundant proteins are generally sampled in such a way that no further information is gained by identifying more peptides; in contrast, low-abundance proteins might be missed completely because the peptides are hidden by the numerous peptides derived from highly abundant proteins. These absent values are often the result of under-sampling, but can also occur by a peptide being present in some samples but not in others, which is perhaps more of a problem. An absent peptide might hold important information but, as it is difficult to distinguish whether missing values are a result of under-sampling or an actual difference in the sample composition, peptides with low coverage are often excluded. Missing values in themselves also decrease the statistical power of any post-analysis. On the contrary, SRM guarantees that data are collected for all defined peptides, resulting in datasets with no or few missing data points. Shotgun, by design, treats all peptides identically, regardless of quantity or other properties making the peptide easier or more difficult to work with in mass spectrometry. In SRM, the acquisition parameters can easily be adapted for a particular peptide to increase the sensitivity or the statistical power by increasing the acquisition time or the number of data points measured. In shotgun, peptides are identified that might have disfavourable properties. As an example, a peptide that is present in more than one protein does not provide much information about the true protein composition of the sample, as finding the peptide can mean that one or both proteins are present. SRM obviously cannot solve the issue of a peptide being part of two proteins, but it easily allows the user to select peptides with favourable properties. SRM, as a consequence, collects more data on a small subset of peptides resulting in significantly smaller datasets with relatively higher information. While this is advantageous, SRM still has several disadvantages. First, it requires deep knowledge about the sample at hand and about which proteins the user would like to quantify. Forgetting to include an important protein at the design stage might invalidate the entire experiment. Second, as specifying a particular measurement time range for each peptide in a so-called scheduled experiment can go wrong, an experiment can completely fail as data might only be collected outside of the actual elution of the peptide. There are many mature and good software solutions for shotgun proteomics which provide users with the ability to select the tools that best fit the current case. SRM, however, lacks the same depth of tools since it is a younger technique, and there is also a higher need for tools. The current set of available tools for SRM can roughly be divided into four sections: experimental design, data collection, data analysis and data interpretation. The rest of the chapter will describe each area in detail and present some of the challenges associated with each area. Fig. 6.1 provides an overview of the SRM workflow and serves as a guide to the rest of the chapter.

Experimental design

Carrying out an SRM experiment for the first time can be overwhelming as the user needs to specify a large amount of parameters before actually collecting any data. Some of these parameters are used to instruct the instrument what data to acquire and when, while other parameters are used in the scoring step to facilitate analysis of the data and to increase the sensitivity of the experiment. In short, the more detailed information you have about what

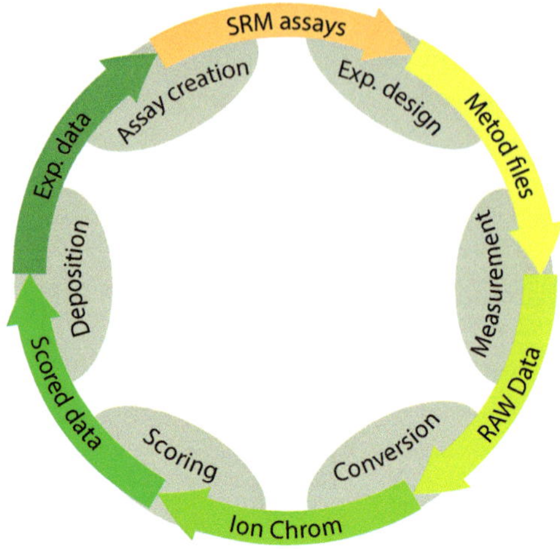

Figure 6.1 Overview of an SRM experiment. A user typically starts at the top of the circle with a set of available SRM assays (section on 'Experimental coordinates: the SRM assay') and then designs the experiment by choosing which SRM assays to measure (section on 'Selection of proteins and peptides'), whether standard peptides used to normalize the retention time are needed (section on 'RT reference peptides to normalize the retention time between experiments') and if heavy reference peptides are required to estimate absolute abundances (section on 'Aqua peptides to estimate the absolute abundance'). Method files, which contain all of the information that the instrument needs to acquire the required data, are created (section on 'Generate instrument method files: using the SRM assays to instruct the instrument'). The instrument produces files in a proprietary format and it is common to convert these files into mzML files, where the data can be stored as ion chromatograms (section on 'Converting data to open formats'). The data are then typically scored (section on 'Estimation of false discovery rates') estimating the false discovery rate. Data are then ideally deposited in one or more data repositories (section on 'Deposition of data into data repositories') and the deposited data can then be used to refine the SRM assay repository so that the next experiment can benefit from the information created by the last experiment. There are several ways to create crude SRM assays if none are available; these techniques are discussed in the sections '*De novo* estimation of an SRM assay', 'Using spectral libraries to guide the creation of SRM assays', 'Online SRM assay repositories: plug-and-play', and 'Using crude synthetic peptides to create SRM assays'.

you want to measure and the sample at hand, the simpler and more powerful SRM becomes. This section describes in detail what is needed and what tools are available to help the user (a list of common tools can be found in Table 6.1).

Selection of proteins and peptides

The user must decide which proteins are of interest. This task varies immensely depending on the hypothesis tested in the current experiment and can be either trivial when working on a well-characterized system where are proteins of importance are known or highly complex when studying systems with less information available. Failing to include a protein of high relevance can only be resolved by re-measuring, at a great expense of both resources and time. As selecting a set of proteins is a more general problem that is relevant to many technologies, it will not be discussed in this chapter; see Malmström *et al.* (2011) and Huang da *et al.* (2009) for a more general discussion. Once a set of proteins has been selected, the

Table 6.1 A list of tools used in the experimental design

Resource	Reference	Section	Note
APEX	Calculating absolute and relative protein abundance from mass spectrometry-based protein expression data (Vogel and Marcotte, 2008)	*De novo* estimation of an SRM assay	Estimate detectability
BiblioSpec	Analysis of peptide MS/MS spectra from large-scale proteomics experiments using spectrum libraries (Frewen *et al.*, 2006)	Experimental coordinates: the SRM assay	Cluster ms2 spectra and generate consensus spectra
DDB	Automated workflow for large-scale selected reaction monitoring experiments (Malmström *et al.*, 2012b)		Web-based tool to design experiments and analyse data.
msClusterer	Clustering millions of tandem mass spectra (Frank *et al.*, 2008)	Experimental coordinates: the SRM assay	Cluster ms2 spectra and generate consensus spectra
PASSEL	PASSEL: the PeptideAtlas SRMexperiment library (Farrah *et al.*, 2012)	Online SRM assay repositories: plug-and-play	Measured SRM data
PeptideART	On the accuracy and limits of peptide fragmentation spectrum prediction (Li *et al.*, 2011)	*De novo* estimation of an SRM assay	Estimate fragment and fragment intensities
PeptideAtlas	PeptideAtlas: a resource for target selection for emerging targeted proteomics workflows (Deutsch *et al.*, 2008)	'Experimental coordinates: the SRM assay'	Peptide seen by mass spectrometry
PeptideSieve	Computational prediction of proteotypic peptides for quantitative proteomics. (Mallick *et al.*, 2007)	'*De novo* estimation of an SRM assay'	Estimate Detectability
Skyline	Skyline: an open source document editor for creating and analysing targeted proteomics experiments (MacLean *et al.*, 2010)		Desktop application to design experiments and analyse data
SpectraST	Development and validation of a spectral library searching method for peptide identification from MS/MS (Lam *et al.*, 2007)	Experimental coordinates: the SRM assay	Cluster ms2 spectra and generate consensus spectra
SRMatlas	PASSEL: the PeptideAtlas SRMexperiment library (Farrah *et al.*, 2012)	Online SRM assay repositories: plug-and-play	TraML files
SSRcalc	Sequence-specific retention calculator. A family of peptide retention time prediction algorithms in reversed-phase HPLC: applicability to various chromatographic conditions and columns (Spicer *et al.*, 2007)	*De novo* estimation of an SRM assay	Estimate relative retention time

user needs to decide which and how many peptides to measure from each protein. Peptide measurability, that is, the relation between the amount of peptide present in the sample and the strength and quality of the measured signal produced, is highly variable; some peptides produce high signals while other peptides fail to reach the limit of detection. The researcher also needs to consider any modifications that affect the peptide as a site with a highly variable modification rate, which negatively influences the quantitation accuracy. All of these properties influence the selection and it is difficult to set definite priorities between the various parameters need to be considered. As such, peptide selection and *de novo* estimation of experimental coordinates is discussed in a single unit, under the *de novo* estimation section. This section is followed by simpler and often more accurate ways of obtaining experimental coordinates. It is a common occurrence that users spike their samples with synthetic peptides to increase the information gained from the experiment; a few of these practices are discussed at the end of the section.

Experimental coordinates: the SRM assay

Before discussing how to select peptides, we need to discuss what information is needed in order to measure and analyse a particular peptide. It goes without saying that the exact amino acid sequence of the peptide to be measured is needed (Malmström *et al.*, 2011). This information can readily be obtained using protein databases to obtain the sequence of the protein and then *in silico* digest the protein into peptides using the same digestive enzyme as in the planned experiment (Cagney *et al.*, 2003). It is of key importance to keep polymorphisms in mind when using generic online databases instead of the genome of the organism in the sample. The reason for this is that online protein databases such as SwissProt (Boeckmann *et al.*, 2003) stores a single sequence out of potentially thousands of variants of the protein present in the population. The result is that a peptide that should be present according to the database might in fact not be present due to these differences and the result would be that that peptide would be deemed absent from the sample at hand. From the peptide sequence, the precursor isolation window (Q1), that is the mass over charge (Thomson, m/z), for a particular charge can be calculated; as peptides fragment mostly along the backbone, the fragment isolation windows (Q3) can also be calculated. The Q1/Q3 combination is called a transition and the relative intensity of transitions belonging to the same assay is important when scoring assays. Knowledge about the retention time is also highly important, but since there is large variation in the retention time between various experimental set-ups, this is normally expressed in a normalized fashion, ntime (Malmström *et al.*, 2012b) and iRT (Escher *et al.*, 2012) being two examples (also see section on 'Normalized retention times'). All of the available information for a particular peptide under a certain charge and modification state is called an SRM assay. The SRM assay is used when designing the experiment, as input for the instrument, and also when analysing and evaluating the data generated. As SRM assays are of key importance to SRM, the Proteomics Standard Initiative has developed a format to capture all of the relevant information; this format is called TraML (Deutsch *et al.*, 2012). Fig. 6.2 provides a graphical reference to an SRM assay and also displays some of the more common strategies in setting the parameters.

De novo estimation of an SRM assay

De novo estimation of the experimental parameters can be performed using the physiochemical properties of the peptide of interest. This is the most generic way, but it is also the most difficult, is less accurate and is rarely used. The software used to estimate these are improved continuously and are becoming more reliable. For a detailed overview, please refer to Reker and Malmström's recent review (Reker and Malmström, 2012). Before estimating the data needed to make an SRM assay, the user needs to pick one of the many peptides that are generated by the digestion. It is common to use peptideSieve (Mallick *et al.*, 2007) or Apex

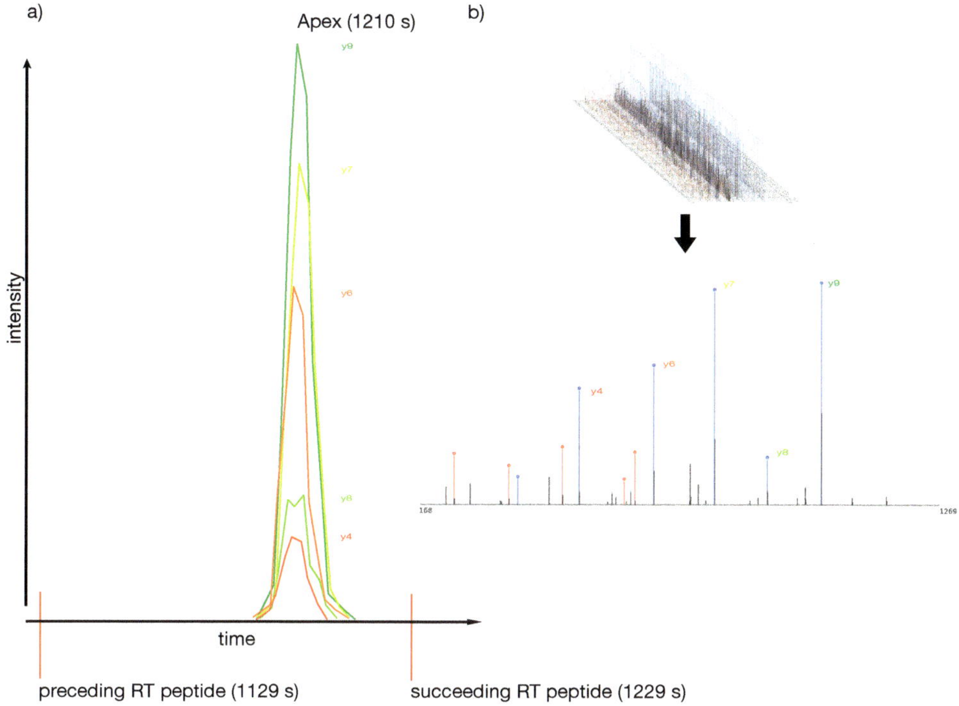

c)

Transition	Q1 M/Z (Z=2)	Q3 M/Z (Z=1)	Relative area SRM	Relative peak height in consensus	Ntime
Y9	629.35	1016.54	1.00	1.00	1210
Y7	629.35	789.41	0.73	0.98	1210
Y6	629.35	660.37	0.51	0.67	1210
Y8	629.35	902.50	0.26	0.21	1210
Y4	629.35	504.28	0.18	0.55	1210

Figure 6.2 Visual display of the data that constitutes the SRM assay. Ion chromatograms for five fragments (y4, y6–y9) from peptide IQNIEGVEVTR at precursor charge 2, fragment charges of 1 are visualized in (a). The jagged elution profiles are a result of the discrete sampling of the mass spectrometer. The apex at 1210s is here reported in normalized retention time, ntime, which is estimated using the preceding and succeeding RT reference peptides eluting at ntime 1129 and 1229, respectively. Ntime is recorded as is the relative area (by integrating the area of each peak) and these are reported in the table in (c). This assay was generated by testing the top 5 y ions in the consensus spectra displayed in (b). The consensus spectra itself was generated from a large number of MS2 spectra. Note that while y4 is the fourth most intense peak in the consensus spectra, y8 is the fourth most intense peak in the SRM data. These small variations make a difference when scoring the data, which is one of the reasons that high-quality SRM assays are best generated on the instrument where the production experiments are running.

(Vogel and Marcotte, 2008) to estimate detectability of the peptides. This score is roughly a combined probability of the enzyme of choice producing the peptide, the peptide ionizing, fragmenting in a favourable way and reaching the detector. It should be mentioned that all of these tools are trained from experimental data generated using shotgun and the assumption is hence that this probability is roughly unchanged for the different experimental techniques. Pilot work has shown that while not perfect, the assumption roughly holds for shotgun and SRM experiments carried out on Thermo instruments (Orbi-trap XL and TSQ quantum). The second piece of information that is often used is SSRcalc (Spicer *et al.*, 2007), which estimates the relative retention time of peptides based on their sequence. This allows the user to place the peptides in a somewhat correct order, but still does not allow the user to estimate real-time elution times without calibrating the transformation between the SSRcalc estimated retention time and the experimental set-up used. The relative fragmentation intensity can be roughly estimated using a tool such as PeptideART (Li *et al.*, 2011). Furthermore, transitions are by no means unique, not even when considering elution times. In other words, the probability of two peptides of roughly the same weight and elution time producing one or more fragments of similar weight is not zero and this probability grows with the complexity of the sample, the inverse of the length of the chromatographic separation and with wider isolation windows. When two signals collide, this is called interference and can be estimated with tools such as SRMcollider (Rost *et al.*, 2012) or ProteomeDB (Sherman *et al.*, 2009). It should be mentioned that estimated values are generally less accurate than experimentally generated data. The hope, of course, is that the techniques used in the estimation process will be refined to the point where the gap closes and the sections below outline some of the tools that can help to close the gap should they be developed or refined.

Using spectral libraries to guide the creation of SRM assays

A quick and excellent option to generate some of the parameters of an SRM assays is to use data from shotgun experiments to make an initial estimate of good assays. Generating consensus spectra from several observed spectra reduces noise and should generally be used. Several tools such as SpectraST (Lam *et al.*, 2007), BiblioSpec (Frewen *et al.*, 2006), msClusterer (Frank *et al.*, 2008) or DDB (Malmström *et al.*, 2002, 2006, 2012b) can be used for this. Several important pieces of information can be extracted from a shotgun library. First, peptides compatible with mass spectrometry can be read directly from the library. The assumption here is that the LC and ionization technologies are similar between the instrument generating the library and the instrument intended to be used for the SRM measurements. Second, the retention time can be estimated and lastly, the most common ways of fragmenting the peptide are recorded (see Fig. 6.2). However, the data are not directly compatible with the different architecture of the instruments commonly used to carry out SRM experiments and hence should ideally be used as a guide. Several tools are available for this, with the most convenient being skyline. Here, a library of BiblioSpec format (Frewen *et al.*, 2006) can be read into the program and the user can conveniently browse through the data selecting peptides and transitions. The challenges here are almost identical to those described in section on '*De novo* estimation of an SRM assay' and will not be discussed further.

Online SRM assay repositories: plug-and-play

Several online resources exist where it is possible to download SRM assays. Some sets are highly specialized and can measure a particular set of proteins, while others are organism-specific, for example *Streptococcus pyogenes* (Karlsson *et al.*, 2012), *Saccharomyces cerevisae*

(Picotti *et al.*, 2013), *Mycobacterium tuberculosis* (Schubert *et al.*, 2013) and others. These libraries constitute a real resource as they can significantly reduce the time invested prior to carrying out an experiment by allowing the user to circumvent a sometimes time consuming SRM assay creation step. More of these libraries are expected to come online as further studies are published and SRM as a technique becomes more established. There are two resources which need to be mentioned explicitly: the first, PASSEL (Farrah *et al.*, 2012), contains several publicly available data files which are processed under a uniform technology making the data explicitly comparable; the second repository is SRMatlas, which is part of the PASSEL technology, but only assays are downloadable in the TraML format.

A major challenge is the portability of experimental coordinates between various instrumental set-ups. When and how to use SRM assays from a different experimental architecture is not a solved problem, but current efforts focus on generating equivalent datasets on various set-ups using the same sample. This will allow for a detailed comparison between datasets. In cases where the experimental set-ups are similar, it might be possible to directly use the SRM assay, but in other cases, this step will result in major challenges. For example, if the fragmentation chamber differs significantly, it is conceivable that different fragments are produced, in which case the downloaded SRM assay will not be of much use. In other cases, the differences might be smaller and it is conceivable that one can develop a set of tools to convert the SRM assays. It is likely that the first tools of this kind will be based on the datasets described above with some machine learning or statistical approaches incorporated.

Using crude synthetic peptides to create SRM assays

There are cases where a particular peptide is of interest for which no experimental data exists because it is from a protein of low abundance. In these cases, one can acquire crude synthetic peptides that can be used to generate experimental data. As crude peptides are relatively cheap, this can be carried out for hundreds of peptides. Many of the assay repositories published as of today have been generated from crude peptides, which implies that most of the parameters will be reasonably accurate. However, not everything can be estimated from crude peptides, with interference being the most obvious example. One way of alleviating this is to report as many transitions for each SRM assay, then run the SRM collider, for example, and simply remove transitions which do not fulfil the criteria.

RT reference peptides to normalize the retention time between experiments

Accurate retention time information plays a key role in many data analysis schemes. The retention time can vary significantly between different experimental set-ups and also the same set-up over time, as components such as the column are regularly exchanged. The variation can be mitigated using a normalized retention time space and, for convenience, people often spike in so-called retention time reference peptides. These peptides are synthetic and are not found in any known natural proteins. They are also optimized to give a strong signal in the mass spectrometer and are designed to span the retention time space of most proteomics experiments. The retention time peptides can be used as reference points, allowing the software to convert the experimental retention times into a normalized retention time space, allowing the direct comparison of measurements carried out on different instrumentation. It also allows for the algorithms to compare the assay retention time to the normalized retention time, as the assay retention times often are expressed in normalized retention time units. While there are commercially available retention time kits, there still might be a good reason to develop a new kit. This kit would ideally have the following key properties: span the entire retention time range, have a minimal impact on the data acquisition, be easily detectable and not interfere with any naturally occurring peptides. It is not possible to fulfil all of these requirements, but reasonable estimates can be accomplished using the tools mentioned above. However, it is still highly recommended that a new reference peptide kit is carefully tested before used in production. The second challenge is true for all reference peptide kits. The fewer peptides, the less interference, but also the cruder the retention time estimates will be. Most kits that are currently in use have less than a dozen peptides. For this reason, most retention time estimates are either based on some linear model or segment based retention time normalization. In shotgun experiments, the informaticians have access to many more anchor points allowing them to correct the retention time using splines, for example. This is far superior in most cases. However, it should be pointed out that the retention time can only be roughly estimated, not definitely determined.

Aqua peptides to estimate the absolute abundance

Targeted proteomics is especially powerful for absolute quantification, as one is guaranteed data from both the sample and the reference of known abundance. The most common references used are the so-called Aqua peptides, which are heavily labelled and hence easily distinguished from the peptides generated from the sample itself. The Aqua peptides are spiked into a sample at known concentrations. The signals from the sample and the heavy reference can then be used to determine the absolute concentration of the sample peptide with high accuracy. However, several considerations need to be made when selecting Aqua peptides. In the easiest case, the user has one Aqua peptide per actual peptide. Here, the user is left to calibrate the amount of Aqua peptide to spike into the sample. The closer this is to the concentration of the endogenous peptide, the more accurately the absolute abundance can be determined. However, this calibration can be tedious if no information on abundance is available, meaning that the user would need to start by spiking the aqua peptides into the solution at their highest concentration and then make any necessary adjustments. Increasing amounts of absolute quantitative data from cell lines or other systems are available, which can be used for an initial estimation. No work in this direction has been published to date, however, which might be a result of Aqua peptides being relatively expensive and their use being relatively restricted. Nevertheless, it is conceivable that such a tool will be developed if the use is increased.

Scheduled experiments

The number of concurrent transitions for which an instrument can acquire data over a given time point has an upper limit; what this upper boundary is depends on several factors, such as dwell time (how much time the instruments is collecting data for a particular transition before switching to the next) and the time difference between two data points in the ion chromatogram. 200 current transitions with a dwell time of 10 ms would result in one measurement every 2 seconds, but since the instrument is using time to switch between the transitions, in practice it takes slightly longer. A user that wants five transitions measured per peptide and two peptides per protein can hence measure 20 proteins per injection. Measuring more proteins requires more than one injection and/or using so-called scheduled measurements. In scheduled experiments, the task is to instruct the instrument to measure a particular peptide only at its expected elution time. This requires two pieces of information: the normalized elution times for all peptides and information on how the normalized elution time relates to the absolute elution time with the current instrument set-up. The user can then define how much time to add before and after the expected elution time. In practice, it is possible to increase the number of transitions to 800 or more, but the exact number of transitions is heavily dependent on the actual set of transitions to be measured. If all peptides elute at the same time, the number of transitions would be lower, whereas if the peptides are evenly spread over the chromatographic elution, the number of transitions would be higher. At times, the only viable option to acquire the desired data are to carry out multiple injections per sample, each with a different set of transitions. Optimally, the peptides are divided between the injections in such a way that allows for the maximum number of transitions to be measured in each injection. Often, however, this means that peptides eluting close to one another are measured in different injections. There are several tools that aim to optimize this, and it is worth using the tool from Bertsch *et al.* (2010), which uses a sophisticated approach to optimally schedule the transitions.

Conclusion

SRM requires significant amounts of prior knowledge in order to be carried out effectively, as discussed in this chapter. There are many tools and sources that can be used to at least obtain an initial estimate of all of the parameters needed and it can be worthwhile iterating a few times to optimize the parameters for a particular experiment. The next chapter will deal with how these parameters are used to guide the instrument to acquire the required data.

Data acquisition and data processing

This chapter aims to introduce how to best instruct the instrument to acquire the data required. This is achieved by generating method files that are copied onto the acquisition computer and then imported into the acquisition software. The resulting data can be stored in a couple of different ways, which are discussed in the subsequent section; finally, the conversion of open formats will be discussed.

Generate instrument method files: using the SRM assays to instruct the instrument

Each instrument uses a method file of proprietary format, which is almost guaranteed to change on a regular basis. This method file needs to contain all of the relevant information present in the SRM assay to allow the instrument to carry out the experiment as efficiently as possible. There are generally two ways of generating method files. The most elegant solution is to use the vendor API to write the files directly. This strategy is used by tools such as skyline (MacLean *et al.*, 2010), but cannot be implemented for tools that do not run on the Windows platform. The less elegant solution is to write text files of a defined format that can be converted to method files by software bundles provided with the instrument. It is relatively easy to generate these files if all of the information is available; therefore, the major challenge lies in the fact that each instrument requires a different format and hence this is a moving target.

Different ways of representing the data

As mentioned in the previous section, data are in the form of ion chromatograms, i.e. two arrays: one specifies the retention time and the other specifies the intensity. Note that there is no m/z information encoded in these chromatograms; instead, this information is stored as metadata for the chromatogram. The logic here is that each informative transition will have a chromatogram and, as the instrument is not scanning, the m/z information is constant. The time dimension is of high value. The reason for this is that the elution of peptides from reverse phase columns is well understood and can be used in data analysis. Additionally, many chromatograms are also time series in that the distance between data points is fixed at, for example, three seconds. As the elution of peptides is comparatively long, one will have several data points that specify the elution profile.

Converting data to open formats

Most vendors rely on proprietary formats to store the acquired data and the tools shipped with the instrument can naturally read these formats, allowing the user to inspect and analyse the data. While many of these tools are highly capable, many users prefer to analyse data using third-party tools. Some of these can read the vendor formats using pre-compiled software from the vendors, which is almost exclusively provided as Windows-only dynamic libraries in the form of DLL files. The alternative is to convert the proprietary formats to open formats such as mzML (Martens *et al.*, 2011). Powerful tools are readily available, with msConvert (Chambers *et al.*, 2012) being one of the more popular ones; msConvert is part of the ProteoWizard software suite (Chambers *et al.*, 2012). While seemingly straightforward, converting the proprietary formats to open formats can at times be challenging. The biggest challenge is the availability of tools that can read the proprietary formats and these are commonly distributed with the instrument in the form of a DLL. The files can obviously not be read unless the user has access to the correct DLL and there are a few cases where backwards-compatibility is less than perfect. This essentially means that it is normally relatively easy to convert a recent file, but it becomes more challenging over time. The second challenge is that data are not always preserved perfectly in the open format. This takes two forms, metadata capture and numerically identical conversion. Not all metadata are captured by the conversion tools; this means that information describing the data is lost in the conversion. This is relatively benign if careful records of all files are stored, but this is not always practised. In that sense, the user loses some of the ability to recover information of metadata after conversion. On a more serious note, it is not always guaranteed that the numerical values of the measurements are absolutely conserved during the conversion. There are several reasons for this, with the most common being that the user can choose to, for example, centroid the data before conversion. As neither the proprietary nor the open formats can capture all of the required information or cases used, it is good practice to store both.

Conclusion

Targeted proteomics requires the user to carefully instruct the instrument on how to acquire the data. The acquired data generally go through a couple of conversion steps to facilitate the usage of open software tools to reduce noise and normalize the data; this is the topic of the next chapter.

Noise reduction and normalization

Measurements always need to be processed in various ways before any analysis can be carried out. Noise is present in the data and there are various distortions and artefacts present that should ideally be removed before analysing the data. If the concentration of one sample is higher than another, it will appear as if all proteins are present at higher amounts in the sample with high concentrations. While this is true, it is generally not the information sought out, and this systematic bias needs to be removed before comparing the samples. In these next paragraphs, the most common processing steps will be discussed.

Noise reduction

Data directly from the instrument can be processed to increase the amount of useful information that can be extracted. There are several operations which are commonly applied, with noise reduction being one of the more prominent ones. Noise reduction is a large research field in itself and hence, only a few principles will be discussed here. The primary objective of noise reduction is to reduce signals that are present in the data which do not come from the peptide of interest. An ever-present source of noise is electronics, especially when the signal is digitally amplified. This is often called white noise and denotes random signals with little or no trends, making it possible to be reduced. The Savitzky–Golay filter is often applied to the chromatograms, in order to smooth the data. Tools such as DDB (Malmström *et al.*, 2012b) rely on wavelets to split the signal into frequency domains and remove signals of high or low frequency, as these signals seldom come from peptides.

Normalized retention times

Another common operation very often used in mass spectrometry is normalization of the retention time. Retention time is a product of several things and varies extensively over time. This means that the same peptide will elute at different times in two experiments (see Fig. 6.3). Note that the order between peptides is more preserved but cannot be relied upon completely. It is common in targeted proteomics to add a set of peptides with highly defined retention times (see section on 'RT reference peptides to normalize the retention time between experiments'), detect these peptides and use them as anchor points to remove the retention time differences between two or more datasets. The most popular retention time correction is based on iRT (Escher *et al.*, 2012), where a linear model is generated from the anchor peptides and then the retention time is aligned with a standard retention time which spans from 1 to 100. Analogous to this is the DDB *ntime,* where the datasets are aligned to an arbitrary retention time scale in a segment-based fashion, where each segment is defined by standard peptides or a set of endogenous peptides if preferred. There are several challenges associated with normalizing the retention time. First, the RT peptides need to be identified and their absolute retention time determined; second, the normalized retention time needs to be estimated for all other peak groups. This is often done by creating a linear model of the retention time using the RT peptides. The assumption here is that the retention time is distorted in a way that can be modelled using a linear function. This is partly alleviated by segment-based retention time models, but these suffer from being relatively accurate in the elution times close to the RT peptides but less accurate further away. More RT peptides result in higher accuracy, but result in the drawback that there is a bigger impact on the sample complexity.

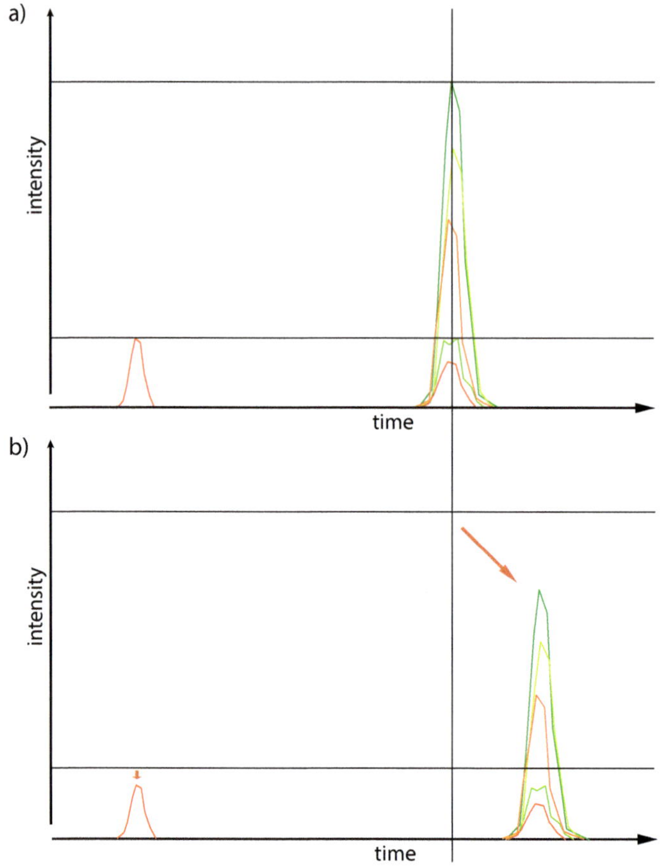

Figure 6.3 Data normalization and RT alignment. Data from two biological replicates are displayed in (a) and (b). The single peak that can be seen to the left in both graphs is an interfering peptide which can easily be discarded as there is no signal for the other transitions. The data from the actual peptide are to the right in the graph with the five transitions each generating co-eluting peaks. Two types of distortions are displayed. The first is the shift in RT that occurred between the interfering peptide and the peptide of interest in (b). The RT shift is best corrected using RT reference peptides, as described in the section on 'Normalized retention times'. The second distortion is in peak height. The difference in height between (a) and (b) can have many explanations, such as there being a real difference between the samples. However, in this case, the difference is due varying concentrations of the samples, where the higher concentration sample in (a) generates larger signals compared to (b); this conclusion is supported by the interfering peptide decreasing by the same amount. The actual difference between the samples in this case was deduced by measuring a large number of peptides from the two samples and observing that the peaks in (b) were systematically lower compared to (a). Normalization of the quantitative values is discussed in the section on 'Estimating interference'.

Normalize the quantification

A major and largely unsolved challenge in targeted proteomics is normalization, where the amplitude of the signal is corrected so that a given intensity in one of the samples in the dataset corresponds to the relative amount of the peptide present in that sample. One very large source of error is the sample concentration and the injection volume, which can easily result in large between-sample variations. In shotgun, it is common to use the total ion current to normalize the dataset; however, this is not possible in targeted proteomics. One strategy used in targeted proteomics is the use of internal standards, i.e. a protein that is believed to be present at a constant amount in the sample. This, in principle, is a good and sound strategy, but it presents challenges when used in practice. The fundamental source of the challenge is that most if not all proteins are regulated under some conditions. Therefore, the challenge often boils down to selecting an appropriate internal standard for the experimental set-up at hand. There is definitely a need for computational tools in this field but exactly which strategy to employ is not clear. One conceivable idea would be to require the user to measure the samples using shotgun first. These data could then be used to identify highly abundant peptide species that do not vary across the samples. These peptides could then be measured in the SRM experiment and used to normalize the intensity. The more normalization peptides used, the more accurate the normalization tool can be. On the other hand, this leaves less instrument capacity to actually measure the peptides of interest.

Estimating interference

One important aspect of targeted proteomics, particularly for instruments with wide isolation windows, is the issue of interfering signals. The source of these signals varies, but one common source is peptides of the same mass which produce fragments of the same mass. In a subset of cases, the peptides also co-elute, in which case the ion chromatogram intensities are the sum of the two co-eluting fragments. It should be obvious that this affects a subset of transitions and that low-abundance proteins are most sensitive to this type of interference compared to high-abundance peptides. Several tools exist that can estimate interference given a genome, with ProteomeDB (Sherman *et al.*, 2009) and SRMcollider (Rost *et al.*, 2012) being two examples. It is also possible to detect interference by comparing the estimated relative abundances between the measured data and the assays. However, this is not foolproof, as the same interference can be present in the assay if it was generated using similar material. Anubis is a tool that automatically estimates interference on the measured data. What makes estimating interference challenging is that it ultimately relies on having an accurate list of all possible ions that can be generated. This is not possible, however, because of the many ways in which fragments can be generated. If all possible neutral losses and internal fragments were considered, the number of fragments would be impractically high. As most of these fragments are very rare, they can normally be excluded.

Conclusion

The raw data need to be processed in several ways before interpretation, as described in this last section. The most obvious variations in the data can be partly resolved with tools and strategies that are currently available, but there is still room for improvement. The next section will discuss how to interpret data.

Data interpretation

The data are now in a format which can be interpreted. The first step is to extract features from the chromatograms and determine which feature belongs to the peptide of interest by scoring all of the features. All features scoring above a user-defined cut off are then used in the downstream analysis and it is commonplace that the error rate is determined. Finally, all features are evaluated against the experiment hypothesis.

Extracting features from the ion chromatograms

Ones the data have been normalized and noise has been reduced, it is time to find the signal that was created by the peptide of interest. This can be trivial for high-abundance and/or well behaved peptides, but becomes more challenging with lower-abundance peptides or those for which there are a lot of interfering signals. Several tools exist to find peaks and estimate their relative intensity. mQuest (Reiter *et al.*, 2011) is an open source Perl-based tool which takes ion chromatograms and assays as input and generates a list of features for each peptide. This is largely done on the command line and does not require much input from the user. Skyline, in contrast, uses similar algorithms but is ultimately dependent on input from the user regarding whether a peak is correct or not. OpenMS is also able to identify peaks in chromatograms and estimate the area of these peaks. As more than one transition is used per peptide, one needs to define which peaks originate from the same peptide. This is often called peak grouping and becomes more challenging with more interfering signals. The most common way to estimate the relative abundance of the peptide is to simply integrate the peak area of all of the peaks in the peak group.

Estimation of false discovery rates

Once the peak groups are defined and their relative intensities have been computed, the challenge is to estimate which peak group is the correct one out of all peak groups found by the particular assay; also, arguably, it is more important to define which peak groups to consider in subsequent steps. An error rate for the selected peak groups is also defined. This is classically a discriminatory analysis and often involves trying to maximally separate a distribution of correct assignments from a distribution of incorrect assignments. mProphet (Reiter *et al.*, 2011) is one tool which uses decoy transitions, e.g. transitions that do not belong to peptides in the sample, to model the negative distribution. mProphet thereafter adapts a scoring function to maximally separate the two distributions and uses the theoretical overlap of the fitted functions to estimate error (see Fig. 6.4). Anubis addresses the problem from a different perspective – it scrambles the data and extracts features from the scrambled data. This distribution will now be the negative distribution. DDB uses a third strategy; here, the ion chromatograms are anonymized in terms of m/z, which allows DDB to compare the measured data with the correct and also unrelated assays, the so-called decoy assays. The score distribution of the decoy assays is then used to estimate the statistical significance of a particular peak group.

Reports and statistical evaluation

Data are now in such a format that the information sought can be extracted, at least in the ideal scenario. One straightforward method is to simply use Student's t-test to estimate the significance of potentially regulated peptides and then use a correction strategy, such as Benjamin-Hochberg, to estimate the number of peptides that have been falsely reported as regulated. More sophisticated strategies can of course be employed; for example, SRM stats allow for optimally utilizing the experimental design. Skyline can use SRMstats (Chang *et al.*, 2012) as an external tool through dropdown menus, making this a convenient option. DDB uses a tool called xplor (Malmström *et al.*, 2012), where expression profiles

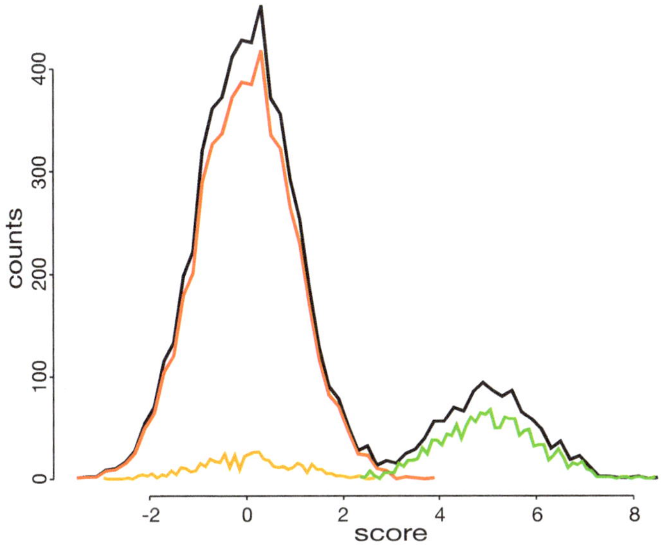

Figure 6.4 The observed score distribution. Each peak group is scored by comparing the data with the SRM assay which was used to generate the data. A better fit between the data and the SRM assay results in a higher score. The observed score distribution (black line) is a result of scoring each peak group, which results in a single correct peak group and many incorrect peak groups at most. The score should separate the correct from incorrect as much as possible, which results in a bi-modal distribution under optimal conditions. The observed distribution is the sum of the incorrect assignments (red line), and correct assignments (green line). The red and green lines need to be estimated as they are unknown. One common strategy is to generate a distribution which is known to be incorrect (yellow line); there are several strategies available to do this. mProphet uses data generated from so-called decoy transitions which do not correspond to peptides believed to be in the sample measured. These decoy transitions can hence be used to identify peak groups which can then be scored. As they cannot be correct, they will be used to estimate the parameters of the negative distribution (red line). All peak groups above a certain cut off, for example 3, will be carried forward at the end of the experiment and the false discovery rate will be determined as the fraction of the red distribution above the score of 3 divided by the green distribution above the score of 3.

are clustered after principal component analysis and proteins are simply grouped without making any statements on whether a particular protein is regulated or not. Which tools or statistical methods to use varies by experiment, but there are several challenges which are not SRM-specific. Comprehensively discussing all of these options is far beyond the scope of this chapter; instead, a single example will be discussed to highlight some of the challenges (see Fig. 6.5). The first choice that needs to be made is whether to analyse each protein individually by using for example a Student's *t*-test, or if the entire dataset should be evaluated together, for example using clustering. The first case is good for picking out strong signals, but suffers from the increasing error rates as more proteins are tested. Evaluating 20 proteins at 95% confidence will statistically result in one false positive. There are different challenges when analysing the dataset as a single unit. One of the key aspects is that data often need to be scaled to obtain a meaningful d result. For example, clustering using the measured intensities will often result in high-abundance proteins clustering together. One way of solving this is to scale the data (van den Berg *et al.*, 2006), which solves the problem

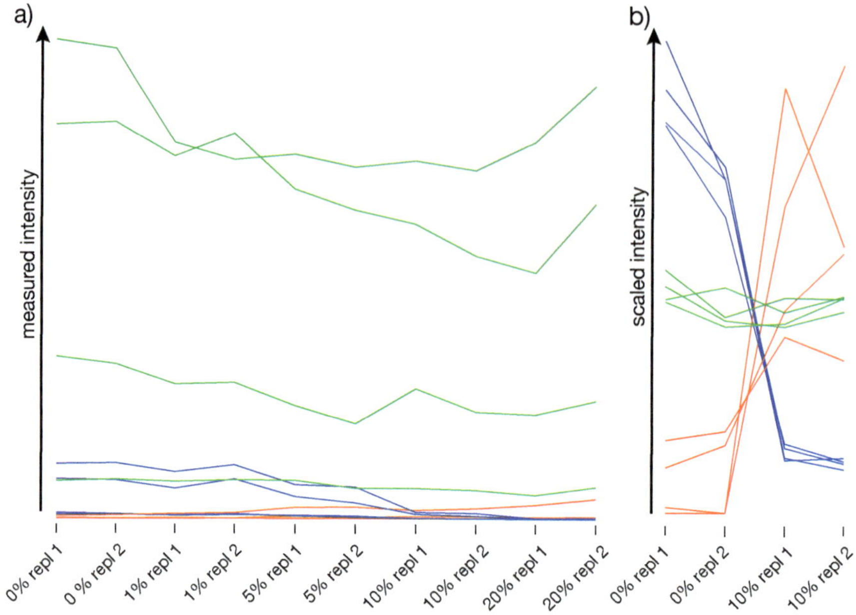

Figure 6.5 Differentially expressed proteins. The samples from Malmström *et al*. (2012a) (*S. pyogenes* grown in 0%, 1%, 5%, 10% and 20% plasma in the growth medium) in biological replicates were measured by SRM. 0% plasma was a proxy for the bacteria living in the oral cavity and growth media with plasma was a proxy for when the bacteria have become invasive and are present in the blood stream. Twelve proteins were measured: four proteins from the Large ribosomal subunit (green lines), four proteins from fatty acid biosynthesis (blue lines) and four proteins believed to be involved in the virulence of the invasive bacteria (red lines). (a) displays the raw data in a parallel coordinate plot, where the raw intensities of each protein are plotted as a line spanning the ten samples. The data in (a) are relatively difficult to analyse as it appears that most proteins are down-regulated. However, this is due to the fact that the same amount of protein was injected in each case and, in higher plasma concentrations, more proteins come from humans and less come from *S. pyogenes*; this results in the appearance of the *S. pyogenes* proteins being down-regulated with increasing plasma. The scale of the different intensities makes it difficult to determine what is going on for the low-abundance proteins. One solution is to analyse each protein independently, but results in less statistical power. In (b), the same data are shown, but here, the data were normalized and scaled in such a way that the intensities of the measurement points were equal for each protein. Also, only data from 0% plasma and 10% plasma are shown. This increases the noise, as signal from low-abundance proteins are noisier and the weight of these proteins is greatly increased. However, the trends are now much easier to visualize.

of the abundance becoming a major factor in the clustering. However, it creates a new problem in that low-abundance proteins have higher relative noise compared to high-abundance proteins. Scaling increases the weight of the low-abundance proteins and invariably results in the noise levels increasing as well (see Fig. 6.5 for an example).

Quality estimation

Acquired data vary in quality depending on a multitude of factors. Important factors are sample quality, the stability of the instrument and the protocols and methods used (Addona *et al*., 2009). No widely adopted standard for estimating the quality for the data has yet been adapted. However, this is an important challenge in the field of targeted proteomics and will likely present some challenges. The corresponding tools for shotgun proteomics often rely on the idea that one would expect signals to be measured continuously and segments with no data are tell tale signs of problems. In targeted proteomics, the absence of signals does not really mean anything, simply due to the fact that one is picking which segments to analyse and that selection can easily produce extended periods of little or no data. Several pieces of information can be extracted from the ion chromatograms that can be used to estimate the quality; signal to noise levels can be estimated, the number of peaks fulfilling certain criteria can be calculated and coherence between the various transitions, or in cases where there are several technical or biological replicas, the consistency of the measurements can be calculated. However, the major hurdle that needs to be cleared is an acceptable definition of what a high quality measurement is. This varies from user to user and therefore cannot be solved on a global level. On the other hand, it would be possible to develop a tool or a set of tools which would estimate these parameters on one or more datasets and then allow the user to manually define which samples he or she considers good quality and which are not. The tool would then build a statistical model to estimate the quality in such a way that the tool could easily be applied to samples not used in the training.

Deposition of data into data repositories

Targeted proteomics is in part powerful due to the re-use of experimental data. It therefore goes without saying that being able to compare the data currently under investigation with prior data are of high importance. This can be accomplished by storing the key features of the experimental data in the form of SRM assays, as described above. However, there are other ways of doing this, and that is where databases that explicitly store all data come into play. While several of these tools have been developed, few have been published to date. There are some, like the SRM atlas, which is operated by a central organization where all of the data will eventually become public. These are important tools which enable scientists to access the data accompanying a publication. There are also others which are inherently designed to store private data, allowing users within the same organization to use the data when analysing a new dataset or when carrying out meta-analyses. There are several challenges associated with this long-term storage. Crucially, the particular assay needs to be carefully annotated, as does information such as acquisition parameters and retention time normalization. These two resources are generated at different times; the assays are designed during method development, while some of the acquisition parameters and retention times are generated during the taking of measurements. The second challenge is in regards to various 'versions' of the SRM assays. It is conceivable that two users developing SRM assays will have slightly different preferences, which can lead to differences between assays despite both users targeting the same peptide. For example, one user might exclude a particular transition due to interference, whereas another did not check for interference. More obvious cases are where different instrument set-ups are used. In these cases, the assays are expected to vary due to differences in fragmentation cell properties, for example. The challenge will hence be related to which information to trust and which to disregard. In most cases, the decision can only be based on careful consideration and also the generation of pilot data to assess variability. Once this variability has been established, one can develop tools to carry forward relevant data, as discussed several times already in this chapter.

Refining the SRM assays from the measured data and scoring the SRM assays

The task of creating high-quality SRM assays becomes easier as the number of data in the repositories grows. The reason for this is that the information needed for the SRM assay can be directly estimated from the experimental data and the main task is simply to minimize the inclusion of noise and incorrect results. Assuming that the data repositories are of high quality and that the amount of incorrect data are minimal, noise can easily be dealt with by simply using the means; the mean will almost always be more accurate compared to any of the individual measurements from which the mean was calculated. It is also possible to identify transitions that do not always behave as expected; these can be excluded, resulting in a better SRM assay. Some data repositories such as DDB also keep track of when an assay was used but did not result in a significant hit, meaning that the signal from the peptide could not be conclusively detected. This can be used to construct an empirical observability score. DDB also keeps track of which samples were analysed, which allows the observability score to be modulated depending on the sample type. Karlsson *et al.* (2012) developed a score according to these ideas. As more abundant proteins always lead to better observability on average compared to low-abundance proteins, the score should mainly be used to pick the optimal SRM assay if more than the desired number of SRM assays is available for the protein of interest.

Conclusion

Scoring the data and filtering out unreliable data are a critical step in SRM. A few tools are currently available, which use different strategies to estimate the false discovery rate at a particular score cut-off. The data passing through the filter are then analysed with the hope that they will be sufficient to address the hypothesis under investigation. Data from the study are then ideally deposited in a data repository, where it can be used in future investigations. For this to be successful, the deposited data need to be annotated in a way that allows for easy navigation and filtering.

Acknowledgement

I would like to thank Johan Malmström for critically reading the manuscript and providing important feedback.

References

Addona, T.A., Abbatiello, S.E., Schilling, B., Skates, S.J., Mani, D.R., Bunk, D.M., Spiegelman, C.H., Zimmerman, L.J., Ham, A.J., Keshishian, H., *et al.* (2009). Multi-site assessment of the precision and reproducibility of multiple reaction monitoring-based measurements of proteins in plasma. Nat. Biotechnol. *27*, 633–641.

van den Berg, R.A., Hoefsloot, H.C., Westerhuis, J.A., Smilde, A.K., and van der Werf, M.J. (2006). Centering, scaling, and transformations: improving the biological information content of metabolomics data. BMC Genomics 7, 142.

Bertsch, A., Jung, S., Zerck, A., Pfeifer, N., Nahnsen, S., Henneges, C., Nordheim, A., and Kohlbacher, O. (2010). Optimal *de novo* design of MRM experiments for rapid assay development in targeted proteomics. J. Proteome Res. *9*, 2696–2704.

Boeckmann, B., Bairoch, A., Apweiler, R., Blatter, M.C., Estreicher, A., Gasteiger, E., Martin, M.J., Michoud, K., O'Donovan, C., Phan, I., *et al.* (2003). The SWISS-PROT protein knowledgebase and its supplement TrEMBL in 2003. Nucleic Acids Re.s *31*, 365–370.

Cagney, G., Amiri, S., Premawaradena, T., Lindo, M., and Emili, A. (2003). In silico proteome analysis to facilitate proteomics experiments using mass spectrometry. Proteome Sci. *1*, 5.

Chambers, M.C., Maclean, B., Burke, R., Amodei, D., Ruderman, D.L., Neumann, S., Gatto, L., Fischer, B., Pratt, B., Egertson, J., *et al.* (2012). A cross-platform toolkit for mass spectrometry and proteomics. Nat. Biotechnol. *30*, 918–920.
Chang, C.Y., Picotti, P., Huttenhain, R., Heinzelmann-Schwarz, V., Jovanovic, M., Aebersold, R., and Vitek, O. (2012). Protein significance analysis in selected reaction monitoring (SRM) measurements. Mol. Cell Proteomics *11*, M111 014662.
Craig, R., and Beavis, R.C. (2003). A method for reducing the time required to match protein sequences with tandem mass spectra. Rapid communications in mass spectrometry: RCM *17*, 2310–2316.
Deutsch, E.W., Lam, H., and Aebersold, R. (2008). PeptideAtlas: a resource for target selection for emerging targeted proteomics workflows. EMBO Rep. *9*, 429–434.
Deutsch, E.W., Chambers, M., Neumann, S., Levander, F., Binz, P.A., Shofstahl, J., Campbell, D.S., Mendoza, L., Ovelleiro, D., Helsens, K., *et al.* (2012). TraML – a standard format for exchange of selected reaction monitoring transition lists. Mol. Cell. Proteomics *11*, R111 015040.
Domon, B., and Aebersold, R. (2006). Mass spectrometry and protein analysis. Science *312*, 212–217.
Escher, C., Reiter, L., MacLean, B., Ossola, R., Herzog, F., Chilton, J., MacCoss, M.J., and Rinner, O. (2012). Using iRT, a normalized retention time for more targeted measurement of peptides. Proteomics *12*, 1111–1121.
Farrah, T., Deutsch, E.W., Kreisberg, R., Sun, Z., Campbell, D.S., Mendoza, L., Kusebauch, U., Brusniak, M.Y., Huttenhain, R., Schiess, R., *et al.* (2012). PASSEL: the PeptideAtlas SRMexperiment library. Proteomics *12*, 1170–1175.
Frank, A.M., Bandeira, N., Shen, Z., Tanner, S., Briggs, S.P., Smith, R.D., and Pevzner, P.A. (2008). Clustering millions of tandem mass spectra. J. Proteome Re.s *7*, 113–122.
Frewen, B.E., Merrihew, G.E., Wu, C.C., Noble, W.S., and MacCoss, M.J. (2006). Analysis of peptide MS/MS spectra from large-scale proteomics experiments using spectrum libraries. Anal. Chem. *78*, 5678–5684.
Geer, L.Y., Markey, S.P., Kowalak, J.A., Wagner, L., Xu, M., Maynard, D.M., Yang, X., Shi, W., and Bryant, S.H. (2004). Open mass spectrometry search algorithm. J. Proteome Res. *3*, 958–964.
Huang da, W., Sherman, B.T., and Lempicki, R.A. (2009). Systematic and integrative analysis of large gene lists using DAVID bioinformatics resources. Nat. Protoc. *4*, 44–57.
Karlsson, C., Malmstrom, L., Aebersold, R., and Malmstrom, J. (2012). Proteome-wide selected reaction monitoring assays for the human pathogen Streptococcus pyogenes. Nat. Commun. *3*, 1301.
Keller, A., Nesvizhskii, A.I., Kolker, E., and Aebersold, R. (2002). Empirical statistical model to estimate the accuracy of peptide identifications made by MS/MS and database search. Anal. Chem. *74*, 5383–5392.
Lam, H., Deutsch, E.W., Eddes, J.S., Eng, J.K., King, N., Stein, S.E., and Aebersold, R. (2007). Development and validation of a spectral library searching method for peptide identification from MS/MS. Proteomics *7*, 655–667.
Lange, V., Malmstrom, J.A., Didion, J., King, N.L., Johansson, B.P., Schafer, J., Rameseder, J., Wong, C.H., Deutsch, E.W., Brusniak, M.Y., *et al.* (2008). Targeted quantitative analysis of *Streptococcus pyogenes* virulence factors by multiple reaction monitoring. Mol. Cell Proteomics *7*, 1489–1500.
Li, S., Arnold, R.J., Tang, H., and Radivojac, P. (2011). On the accuracy and limits of peptide fragmentation spectrum prediction. Anal. Chem. *83*, 790–796.
MacLean, B., Tomazela, D.M., Shulman, N., Chambers, M., Finney, G.L., Frewen, B., Kern, R., Tabb, D.L., Liebler, D.C., and MacCoss, M.J. (2010). Skyline: an open source document editor for creating and analyzing targeted proteomics experiments. Bioinformatics *26*, 966–968.
Mallick, P., Schirle, M., Chen, S.S., Flory, M.R., Lee, H., Martin, D., Ranish, J., Raught, B., Schmitt, R., Werner, T., *et al.* (2007). Computational prediction of proteotypic peptides for quantitative proteomics. Nat. Biotechnol. *25*, 125–131.
Malmström, J., Karlsson, C., Nordenfelt, P., Ossola, R., Weisser, H., Quandt, A., Hansson, K., Aebersold, R., Malmstrom, L., and Bjorck, L. (2012a). Streptococcus pyogenes in human plasma: adaptive mechanisms analyzed by mass spectrometry-based proteomics. J. Biol. Chem. *287*, 1415–1425.
Malmström, L., Malmström, J., Marko-Varga, G., and Westergren-Thorsson, G. (2002). Proteomic 2DE database for spot selection, automated annotation, and data analysis. J Proteome Res. *1*, 135–138.
Malmström, L., Marko-Varga, G., Westergren-Thorsson, G., Laurell, T., and Malmstrom, J. (2006). 2DDB – a bioinformatics solution for analysis of quantitative proteomics data. BMC Bioinformatics *7*, 158.
Malmström, L., Malmström, J., and Aebersold, R. (2011). Quantitative proteomics of microbes: Principles and applications to virulence. Proteomics *11*, 2947–2956.
Malmström, L., Malmström, J., Selevsek, N., Rosenberger, G., and Aebersold, R. (2012b). Automated workflow for large-scale selected reaction monitoring experiments. J. Proteome Res. *11*, 1644–1653.

Malmström, L., Nordenfelt, P., and Malmström, J. (2012c). Business intelligence strategies enables rapid analysis of quantitative proteomics data. J. Proteome Sci. Comput. Biol. *1*, 5.

Martens, L., Chambers, M., Sturm, M., Kessner, D., Levander, F., Shofstahl, J., Tang, W.H., Rompp, A., Neumann, S., Pizarro, A.D., *et al.* (2011). mzML – a community standard for mass spectrometry data. Mol. Cell. Proteomics *10*, R110 000133.

Nesvizhskii, A.I., Keller, A., Kolker, E., and Aebersold, R. (2003). A statistical model for identifying proteins by tandem mass spectrometry. Anal. Chem. *75*, 4646–4658.

Olsen, J.V., Schwartz, J.C., Griep-Raming, J., Nielsen, M.L., Damoc, E., Denisov, E., Lange, O., Remes, P., Taylor, D., Splendore, M., *et al.* (2009). A dual pressure linear ion trap Orbitrap instrument with very high sequencing speed. Mol. Cell. Proteomics *8*, 2759–2769.

Perkins, D.N., Pappin, D.J., Creasy, D.M., and Cottrell, J.S. (1999). Probability-based protein identification by searching sequence databases using mass spectrometry data. ELECTROPHORESIS *20*, 3551–3567.

Picotti, P., Clement-Ziza, M., Lam, H., Campbell, D.S., Schmidt, A., Deutsch, E.W., Rost, H., Sun, Z., Rinner, O., Reiter, L., *et al.* (2013). A complete mass-spectrometric map of the yeast proteome applied to quantitative trait analysis. Nature *494*, 266–270.

Reiter, L., Rinner, O., Picotti, P., Huttenhain, R., Beck, M., Brusniak, M.Y., Hengartner, M.O., and Aebersold, R. (2011). mProphet: automated data processing and statistical validation for large-scale SRM experiments. Nat. Methods *8*, 430–435.

Reker, D., and Malmstrom, L. (2012). Bioinformatic challenges in targeted proteomics. J. Proteome Res. *11*, 4393–4402.

Rost, H., Malmstrom, L., and Aebersold, R. (2012). A computational tool to detect and avoid redundancy in selected reaction monitoring. Mol. Cell Proteomics *11*, 540–549.

Schubert, O.T., Mouritsen, J., Ludwig, C., Rost, H.L., Rosenberger, G., Arthur, P.K., Claassen, M., Campbell, D.S., Sun, Z., Farrah, T., *et al.* (2013). The Mtb proteome library: a resource of assays to quantify the complete proteome of *Mycobacterium tuberculosis*. Cell Host Microbe *13*, 602–612.

Sherman, J., McKay, M.J., Ashman, K., and Molloy, M.P. (2009). Unique ion signature mass spectrometry, a deterministic method to assign peptide identity. Mol. Cell. Proteomics *8*, 2051–2062.

Spicer, V., Yamchuk, A., Cortens, J., Sousa, S., Ens, W., Standing, K.G., Wilkins, J.A., and Krokhin, O.V. (2007). Sequence-specific retention calculator. A family of peptide retention time prediction algorithms in reversed-phase HPLC: applicability to various chromatographic conditions and columns. Anal Chem *79*, 8762–8768.

Sturm, M., Bertsch, A., Gropl, C., Hildebrandt, A., Hussong, R., Lange, E., Pfeifer, N., Schulz-Trieglaff, O., Zerck, A., Reinert, K., *et al.* (2008). OpenMS – an open-source software framework for mass spectrometry. BMC Bioinformatics *9*, 163.

Weisser, H., Nahnsen, S., Grossmann, J., Nilse, L., Quandt, A., Brauer, H., Sturm, M., Kenar, E., Kohlbacher, O., Aebersold, R., *et al.* (2013). An automated pipeline for high-throughput label-free quantitative proteomics. J. Proteome Res. *12*, 1628–1644.

Vogel, C., and Marcotte, E.M. (2008). Calculating absolute and relative protein abundance from mass spectrometry-based protein expression data. Nat. Protoc. *3*, 1444–1451.

Yates, J.R., 3rd, Eng, J.K., McCormack, A.L., and Schieltz, D. (1995). Method to correlate tandem mass spectra of modified peptides to amino acid sequences in the protein database. Anal. Chem. *67*, 1426–1436.

Standardized Formats, Report Information Guidelines, Mass Spectrometry-based Repositories and Application Programming Interfaces for Implementing Data Standards in Proteomics

7

J. Alberto Medina-Aunon and Juan P. Albar

Abstract

Both data standards and public databases or repositories are key factors for the development of the existing and new 'omics' approaches. Throughout this book chapter, we first review the current status of the different proteomic data standards developed by the Proteomics Standards Initiative of the Human Proteome Organization (HUPO-PSI). Secondly, the main existing mass spectrometry (MS)-based proteomics resources PRIDE, Peptide Atlas, Human Protein Atlas, Global Protein Machine Database and the ProteoRed MIAPE Database are reported. Then, we describe how to improve the integration of data standards and databases into the laboratory workflows through the different Application Programming Interfaces (APIs) and existing bioinformatics tools. This section covers the current trends for both bioinformatics developers and non-expert users.

Finally, we review the status of available bioinformatics tools to convert raw data into proteomic data standards as well as data analysis tools for inspection, validation, reporting and statistics.

Background: HUPO-PSI

In proteomics research, the historic heterogeneity of data formats provided by the equipment vendors has hindered progress and made it more challenging to verify the results of published studies. Share and describe proteomics experiments with sufficient detail to enable replication of results are fundamental principles of scientific research that may be paradigmatic in proteomics. Early works from outstanding research groups and consortia have aimed to define public repositories for proteomics datasets with agreed formats to store and exchange the analytical information derived from proteomics labs (Taylor *et al.*, 2003;

Carr *et al.*, 2004; Garwood *et al.*, 2004; Pedrioli *et al.*, 2004). Specifically, in the field of the '-omics' sciences, data standards and standardized use of terminologies and ontologies for biomedical informatics play an important role in reporting high-throughput experiment results in formats that can be interpreted by both researchers and analytical tools. (Deus *et al.*, 2012)

The initiative on standardization within the Human Proteome Organization HUPO-PSI has now been working for more than ten years (Orchard *et al.*, 2012) to improve the standardization of data, drawing together academic and industrial partners in an open collaboration. The stated aims of the HUPO-PSI are to 'define community standards for data representation in proteomics to facilitate data comparison, exchange and verification' (Kaiser, 2002; Orchard *et al.*, 2003). From its foundation in 2002, PSI has held annual meetings, as well as more frequent workshops, that have contributed to numerous improvements to deal with its motto.

The main organizational units of the HUPO-PSI are the workgroups devoted to Molecular Interactions (MI), Mass Spectrometry (MS), Proteomics Informatics (PI), Protein Modifications (MOD), and Protein Separation (PS) (http://www.hupo.org/research/psi/). Each workgroup has the following document outputs:

1. Modules for minimum reporting guidelines.
2. Formal exchange formats usually represented in Extensible Markup Language (XML) for communicating between software packages or sending results to public databases. They normally capture experimental results as well as significant additional details about how the experiments were performed.
3. Semantic web validation by controlled vocabularies (CVs) to provide a standard terminology for the data elements represented within data formats.

To deal with the first of these deliverables, in 2007, the HUPO-PSI published a collective paper to define the Minimum Information About a Proteomics Experiment (MIAPE) specifications (Taylor *et al.*, 2007). Following the root document, a set of MIAPE documents for different proteomics techniques was also delivered. Each MIAPE module collects a minimal checklist of items that should be reported for the given technique. The items were stated using plain language and describe specific points related to the experimental protocol and also to the data analysis workflow. These descriptors, usually called metadata should guide other groups to interpret the published results without ambiguity as well as to how they were generated. Once these guidelines reached a recognized level of maturity, some of the most relevant proteomics journals (Orchard and Ping, 2009; Orchard *et al.*, 2009) encouraged submitters to meet these guidelines to ensure both quality control and data reproducibility in novel proteomics works. In relation to the rest of the deliverables, the HUPO-PSI has also developed data exchange formats and controlled vocabularies applicable to each experimental module, making data sharing and reporting less time-consuming for both proteomics experts and occasional users.

Standardization pillars: MIAPE guidelines, XML-based standard formats and controlled vocabularies

Reporting guidelines: minimal information about a proteomic experiment (MIAPE documents)

The first MIAPE document comprises the essential elements to be reported within a mass spectrometry analysis. This MIAPE–MS (Taylor *et al.*, 2008) is divided along specific sections regarding to the mass spectrometer configuration (ion source, analyser and detector) and to the generation of the resulting peak list.

A second document, the MIAPE-MSI (Binz *et al.*, 2008) is focused on the mass spectrometry-based peptide and protein identification and characterization steps of the proteomics experiment. Data regarding to input data (MS data), searching parameters (database, tolerances, endo-proteases, etc.) and the results, including identified proteins and peptides as well as their interpretation and validation, should be included through this report.

Other documents devoted to protein and peptide separation techniques such us the MIAPE guidelines for gel-based experiments MIAPE-GE (Gibson *et al.*, 2008), gel-based image analysis MIAPE-GI (Hoogland *et al.*, 2010), column chromatography MIAPE-CC (Jones *et al.*, 2010) and capillary electrophoresis MIAPE-CE (Domann *et al.*, 2010) have also been published.

Finally, the most recent MIAPE document, MIAPE-Quant (Martinez-Bartolome *et al.*, 2013), describes a wide range of MS-based quantitative approaches, including labelled, targeted and label-free techniques. In detail, a MIAPE-Quant report contains the experimental design and the description of the starting biological material and conditions, including samples, assays descriptors, group structure (resulting from experimental conditions and their values) and replicate structure. Additionally, a reference to the quantitative protocol and the resulting data could be also included.

The extensible markup language (XML) to standardize proteomics data

In the context of a typical proteomics LC-MS based workflow, the initial step at the level of mass spectrometry is to capture unprocessed and processed spectra (e.g. peak lists for peptide/protein identification). For this purpose the HUPO-PSI XML data standard, called mzML (Martens *et al.*, 2011), allocates the relevant information related to the MS acquisition step in a basic XML architecture. In addition, for identification by selected reaction monitoring analysis (SRM), the TraML standard format (Deutsch *et al.*, 2012) captures the input parameters at the first quadrupole analyser as well as the data transition data from the mass spectrometer.

On the other hand, the standardized format mzIdentML (Jones *et al.*, 2012) contains the output of peptide/protein identification software generated by the database search engine.

The HUPO-PSI has recently released two other standard formats for capturing the output from quantitation software: the mzQuantML (Walzer *et al.*, 2013), an XML format capturing a detailed evidence trail, and the mzTab (https://code.google.com/p/mztab/), a tabular-separated format capturing a simple summary of results.

In addition, the HUPO-PSI has also released the GelML standard format (Gibson *et al.*, 2010) for describing gel-based results. Lastly, the HUPO-PSI is also very active in the standardization of protein interaction data (Hermjakob *et al.*, 2004), although work on interactions is out of scope for this chapter.

The semantic validation: controlled vocabularies

The HUPO-PSI Controlled Vocabulary documents (CVs) provides a consensus annotation system to standardize the meaning, syntax and formalism of terms used across proteomics experiments at all stages. Each PSI working group have developed the CVs terms required by the technology or data type following common recommendations for their development and maintenance. In addition, a common mapping schema has been proposed to describe the exchange associations among its specific elements and the rest of PSI CVs or other ontologies.

Table 7.1 summarizes the standardization pillar documents delivered by the HUPO-PSI group.

Table 7.1 Development of MIAPE modules, data exchange formats and CVs

Techniques	Reporting guidelines (MIAPE)	Data Exchange Format (XML)*	Cont. Voc. (CV)
Mass spectrometry	Mass spectrometry (MIAPE-MS) v. 2.98	mzData (depreciated) mzML v. 1.1.0 traML (SRM/MRM) v. 1.0.0	PSI-MS v. 3.56.0
	Identifications (MIAPE-MSI) v. 1.1	mzIdentML v. 1.1.0	
	MS Quantitation (MIAPE-Quant) v. 1.0	mzQuantML v. 1.0.0 mzTab v. 1.0.0	
Protein separation	Gel electrophoresis (MIAPE-GE) v. 1.4	GelML v. 1.1.0	SEP-CV v. 1.0.0
	Gel informatics (MIAPE-GI) v. 1.0	None	
Sample processing	Column chromatography (MIAPE-CC) v. 1.1	spML (depreciated)	
	Capillary electrophoresis (MIAPE-CE) v. 0.93		

* Available at http://www.psidev.info/

Proteomics repositories and databases

The MIAPE guidelines, the PSI XML data formats and the controlled vocabularies define the way for data reporting in a structured and standardized manner. Once data and metadata are generated using these community standards they usually are exported to some databases or repositories for data sharing, storing and proper spreading and publication. Data deposition on public repositories is one of the final stages in the laboratory pipeline. Sending the experimental results and metadata to proteomic public repositories with sufficient detail to enable further reanalysis of the data are a fundamental tenet of scientific research and may allow new experimental approaches for complementing the initial hypothesis. Here, we describe some of the most popular public repositories to send and share experimental proteomics results.

Protein identification database – PRIDE (http://www.ebi.ac.uk/pride/)

Following the initial steps of HUPO-PSI, the EMBL-EBI (Hinxton, UK) developed a centralized public data repository (PRIDE) focussing on protein and peptide identifications (Vizcaino *et al.*, 2013; Martens *et al.*, 2005). PRIDE stores mass spectra (peak lists), peptide and protein identification data and as much their associated metadata as possible. Along the last years, there has been a substantial increase in the amount of stored data in PRIDE (Table 7.2), becoming the proteomics data submission point for several key journals in the field.

Prior to the establishment of the current stable standards mzML and mzIdentML formats, PRIDE developed another XML schema, the PRIDE XML as an essential complementary element to the main public repository. It describes how data should be represented and uploaded into PRIDE. This widely used XML-based format has been accepted as a data standard for reporting MS and MSI data within a single file.

Table 7.2 PRIDE repository growing

	September 2009	September 2012	November 2013
MS-based experiments	9908	25,853	30,176
Identified proteins (approx.)	2,500,000	11,000,000	1,442,5000
Identified peptides (approx.)	11,500,000	61,900,000	83,800,000
MS Spectra	50,300,000	324,000,000	411,600,000

PeptideAtlas (http://www.peptideatlas.org/)

PeptideAtlas is a multi-organism repository and publicly accessible compendium of identified peptides provided by large sets of tandem mass spectrometry data (Deutsch, 2010; Desiere *et al.*, 2006). Developed by the Institute for Systems Biology (Seattle, WA, USA), mass spectra files for human, mouse, yeast, and several other organisms were collected and reprocessed using the latest versions of the proteomics search engine software and protein sequence databases. The key feature of this repository is all results are subsequently reprocessed through the Trans Proteomic Pipeline (Deutsch *et al.*, 2010). It offers a uniform manner to derive a probability of correct identifications for all results, focussing in obtaining low false discovery rates, thus providing an accurate and up-to-date vision of a particular proteome. The identified peptide sequences are then mapped onto their respective genome/proteome sequences, By November 2013, more than 1000 experiments, containing the raw data, search engine results, and full builds can also be downloaded for other uses. About data statistics, the last reference in 2010 (Deutsch) pointed out around 90,000,000 of MS spectra, 10,000,000 peptides and 42,500 distinct proteins.

In addition to the traditional MS repository, the PeptideAtlas portal also supports data from Selected Reaction Monitoring (SRM) experiments. Therefore, this sub repository called SRMAtlas (http://www.srmatlas.org) is a compendium of targeted proteomics assays to detect and quantify proteins in complex proteome digests.

The global protein machine database – GPM/GPMDB (http://www.thegpm.org/)

The global proteome machine database (GPMDB) is the proteomics repository with the highest number of data, including spectra, peptide and protein identifications. By October 2013 it included 238,000 models (experiments), and more than 127,000,000 and 1,029,000,000 proteins and peptides respectively. Originally it was developed by Beavis Informatics Ltd (Canada) and was constructed to utilize the information obtained by GPM servers to aid in the difficult process of validating peptide sequences from MS/MS spectra as well as protein coverage patterns. This database has been integrated into the GPM server pages, allowing users to quickly compare their experimental results with the best results that have been previously observed by other users.

The GPMDB was designed to be a simplification and extension of the MIAPE scheme proposed by the PSI committee of HUPO. Rather than being a complete record of a proteomics experiment, this database holds the minimum amount of information needed for certain bioinformatics-related tasks such as sequence assignment validation. Furthermore, GPM also supports proteomics targeted experiments using SRM.

ProteomeXchange (http://www.proteomexchange.org)

The ProteomeXchange consortium has been set up to provide a coordinated submission portal of mass spectrometry-related data to the main existing proteomics repositories, including the previously described PRIDE, PeptideAtlas and GPMDB. The capabilities of the different resources are somehow redundant but also complementary, so the idea behind ProteomeXchange is to have a single point of submission for data deposition, while providing multiple points of access for data visualization and analysis. ProteomeXchange defines two pipelines depending on the nature of the data. Thus MS/MS data are processed and checked using PRIDE-based pipeline and SRM/MRM data are processed through PASSEL (PeptideAtlas SRM Experiment Library) (Farrah *et al.*, 2012) framework. A complete list of the public experiments can be obtained and downloaded at http://proteomecentral.proteomexchange.org/. By October 2013, 467 projects were successfully deposited (93% by PRIDE and 7% by PASSEL).

The ProteoRed MIAPE database (PMDB) (http://www.proteored.org/miape)

The ProteoRed MIAPE Database (Martinez-Bartolome *et al.*, 2010) (PMDB) stores the proteomics experimental information following strictly the MIAPE guidelines for gel electrophoresis, gel image analysis, MS and MSI experiments. The tool guides users through each stage of an experimental process, capturing key details as specified in each MIAPE module. Most used controlled vocabularies have been implemented to ensure that consistent method descriptions and units are provided throughout.

This repository offers a user-friendly graphical interface where any user could navigate through the hierarchical structure comprised by the original sections of a MIAPE document. It is always linked to a project and each project can contain an unlimited number of MIAPE reports. Currently, the MIAPE generator tool is able to create and manage gel electrophoresis, gel informatics, mass spectrometry and mass spectrometry informatics MIAPE documents.

By November 2013, PMDB contained 520 proteomics projects with more than 3500 MS-based experiments, including 7500 MIAPE reports (around 400 Gel-based) and approximately 1,286,000 and 37,520,000 identified proteins and peptides respectively.

All the described repositories and summarized in Table 7.3.

Table 7.3 Main MS-based repositories linked with data standards

	PRIDE 11/2013	PeptideAtlas 2010	GPMDB 10/2013	ProteoRed DB 11/2013
MS-based experiments	30,176	1000	238,000	3500
Identified proteins (approx.)	14,425,000	42,450 (distinct)	127,000,000	1,286,649
Identified peptides (approx.)	83,800,000	10,000,000	1,029,000,000	37,518,438
MS spectra	41,600,000	90,000,000	NA	NA
MS-MIAPE reports	NA	NA	NA	7500

Application Programming Interfaces (APIs) for the integration of data standards into laboratory pipelines

Many popular commercial or open source proteomics software packages exist before the HUPO-PSI standards were proposed and thus, they generated their own MS file formats for data reporting and storing. As the HUPO-PSI standards have become more accepted and established by the proteomics community for data reporting, these packages have either already adopted the new standards, or are on their way for supporting them. New software tools for proteomics are also adopting the HUPO-PSI standards. In-house software development in many proteomics labs also needs to deal with the new standards on a day-to-day basis. This requires open-source application programming interfaces (APIs) that strictly adhere to PSI standards while remaining memory-efficient and easy to use. In the following section, we will describe some open source APIs developed to support the HUPO-PSI standards at the various levels of data reporting. Table 7.4 summarizes the APIs reported in this chapter.

Table 7.4 Summary of data standards APIs

	Standard/format(s)	Link
jmzML	MS. mzML	http://code.google.com/p/jmzml/
jmzIdentML	MSI. mzIdentML	http://code.google.com/p/jmzidentml/
jmzReader	MS. Usual PeakLists	http://code.google.com/p/jmzreader/
jmzQuantML	Quant. mzQuantML	https://code.google.com/p/jmzquantml/
jTraML	MS/SRM. traML	http://iomics.ugent.be/jtraml
PRIDE XML	MS/MSI. PRIDE	http://code.google.com/p/pride-toolsuite/
MIAPE API	GE/GI/MS/MSI. MIAPE	http://www.proteored.org/MIAPE/

The Java mzML API: jmzML

jmzML (Cote *et al.*, 2010b) is a Java API for supporting the mzML file format. Typical mzML files shows two main drawbacks for their daily use. First, they can be several gigabytes in size. Second, the mzML format uses abundant referencing among its constituent XML elements. Both disadvantages make it difficult to handle and to parse the whole file in memory. jmzML applies an intelligent XPath[1]-based XML indexing mechanism to circumvent both the described problems of file size and the reference resolution. The XML indexer, known as *xxindex*, is an integral part of jmzML and plays an important role in the other APIs described below. It first indexes a file by noting the start and end byte locations of the XML elements in the file, and then subsequently allows the user to recover the element by specifying its XPath. This mechanism allows the whole mzML file to be treated like a swap-file, where the required elements can be *looked-up* in an on-demand basis, rather than loading the whole file in memory. The jmzML API also comes with a powerful interactive mzML viewer that can be used to simultaneously load and view spectra and chromatograms from multiple large files.

The Java mzIdenML API: jmzIdentML

jmzIdentML (Reisinger *et al.*, 2012) is an interface for the mzIdentML 1.1 standard for peptide and protein identification. It has a similar architecture to jmzML, being built using the xxindex. Furthermore, it presents advanced features for resolving the multiple levels of in-depth referencing built into the mzIdentML format including the protein inference. The API is highly efficient in its memory usage and provides capabilities to the end user for reading and writing mzIdentML documents.

The Java MS data reader API: jmzReader

jmzReader (Griss *et al.*, 2012) is a library of Java APIs to parse MS peak list files. It supports DTA, Mascot general file (mgf), ms2, mzData, mzXML, PKL, mzML and PRIDE XML formats. Six format-specific parsers constitute the library and they have been included into a common interface to provide a simple set of functions to access the core information in any peak list file. All the parsers in jmzReader are fully optimized to support integration with the mzidentML format. mzIdentML files reference spectra in external peak list files, but do not contain information on those spectra. However, jmzReader provides robust support for integrating information in mzIdentML files with the corresponding information in peak list. Also it provides a graphical user interface (GUI), jmzReader-GUI, to support visualization of supported peak-list files.

The java TraML and mzQuantML API: jTraML and jmzQuantML

jTraML (Helsens *et al.*, 2011) provides a Java implementation of the PSI's TraML, which serves as an input format for the parent ion selection and MS/MS transitions in SRM experiments. jTraML also supports some vendor specific formats from ABSciex, Thermo Scientific and Agilent Technologies.

jmzQuantML is a Java API used for reading and writing mzQuantML files recording quantitative proteomic data and meta-data. Together the API, some examples are also available in Google code site that can illustrate users with the most important features.

The PRIDE API: PRIDE XML JAXB

The PRIDE XML JAXB (Java Architecture for XML Binding) library was developed exclusively to parse PRIDE XML 2.1 files (for rapid and memory-efficient reading of PRIDE

1 XPath is a formal description of one complete branch within the tree structure in XML.

XML files). It is a lightweight implementation of the full PRIDE XML 2.1 format that uses *xxindex* indexing for fast access and a JAXB based reference-resolving mechanism to give direct access to the object models of the entities present in the PRIDE XML file.

The MIAPE API

This Java API was developed to automate the generation of MIAPE documents. Using this API, third-party applications will be able to create, store and delete MIAPE documents using either MIAPE central repository or a local build. The project is multimodular, preventing unnecessary dependencies, and improving extension, maintenance and reuse capabilities.

Bioinformatics tools for creating and managing proteomics data standards

Translating MS raw data into mzML and traML standardized formats

Starting from the MS data acquisition process, we have to define how to convert the data provided by the mass spectrometer's software into mzML standard format. Even though there are numerous free and commercial software packages that support mzML formats (http://www.psidev.info/mzml) some of them do not provide a direct method for exporting the acquired MS data as mzML. For these packages, the use of additional translators and transformation utilities is required. The main ones are summarized below (Table 7.5).

Table 7.5 Create and manage data standards

Tool	To get	Comments	Link
ProteoWizard/ MSConvert	mzML	Supports Thermo RAW, Agilent MassHunter.d Bruker compass, FID, YEP and BAF, Waters MassLynx RAW and AB Sciex wiff formats	http://proteowizard.sourceforge.net/
OpenMS	mzML, mzIdentML	It integrates Mascot, OMSSA and X!Tandem output	http://open-ms.sourceforge.net/
ATAQS	mzML, traML	It produces an automated way of generating all verified transitions in TraML	http://tools.proteomecenter.org/ATAQS/ATAQS.html
Mascot	mzIdentML	Results are directly exported to mzIdentML. Supports mzML as input format	http://www.matrixscience.com/
ProCon	mzIdentML	Support conversion from Sequest.out files, Proteome Discoverer® to mzIdentML Also ProteinScape® to mzIdentML and PRIDE XML	www.medizinisches-proteom-centre.de/ProCon
mzIdentML parsers	mzIdentML	Parsers for converting the OMSSA and X!Tandem output formats into mzIdentML	http://code.google.com/p/mzidentml-parsers/
PRIDE converter 2	PRIDE XML	Converts mass spectrometry data from common data formats into PRIDE XML	https://code.google.com/p/pride-converter-2/
ProteoRed MIAPE Webtoolkit	MIAPE and PRIDE XML	It provides bidirectional conversion between MIAPE and XML reports. Full mzML, mzIdentML and PRIDE XML support	http://www.proteored.org/MIAPE/

ProteoWizard
(http://proteowizard.sourceforge.net/)

One of the most robust solutions to facilitate integration of standard formats into lab workflows is ProteoWizard (Kessner *et al.*, 2008). This framework contains a set of modular and extensible open-source, cross-platform tools and software libraries that facilitate proteomics data analysis and perform standard chemistry and LC-MS dataset computations. Currently, in the Windows version using the MSConvert (or MSConvertGUI) tool it is possible to translate raw data formats from Thermo, Agilent, Bruker, Waters and AB Sciex MS instruments into mzML standard format. The tool may also produce MS/MS (MS2) peak lists as inputs for search engines, either expressed in mzML format or in other common formats, such as mgf. Proteowizard should be stable and robust in the future, since it uses software libraries developed by the mass spectrometry vendors and thus if a vendor's raw format changes, a new library should be released and integrated into Proteowizard.

OpenMS
(http://open-ms.sourceforge.net/)

Another solution for end users is OpenMS (Kohlbacher *et al.*, 2007, Bertsch *et al.*, 2011). This is an open-source software for LC-MS data management and analysis. It offers an infrastructure for the development of MS analysis software, using internally the PSI's mzML and mzIdentML standards. It provides utilities for signal processing, allowing baseline reduction and noise filtering of raw data and quantification by most popular methods. In addition, OpenMS can read and write different search engine data formats such as Mascot, OMSSA and X!Tandem, and supports the new PSI quantitation standardized formats like mzQuantML and mzTab.

ATAQS
(https://www.systemsbiology.org/ataqs-computational-software-tool-high-throughput-transition-optimization-and-validation-selected)

The Automated Targeted And Quantitative System (ATAQS) (Brusniak *et al.*, 2011) is a web-based collaborative system for managing an entire targeted proteomics workflow from the beginning (protein/peptide and transition selection) to the end (quantitative analysis of the chromatograms). Transition lists may be imported, stored, manipulated, and exported using several formats, including TraML. Thus, ATAQS is an open-source computational software for high throughput transition optimization and validation for selected reaction monitoring mass spectrometry. ATAQS is the first open-source software that supports all steps of the targeted proteomics workflow. ATAQS also provides software API (Application Program Interface) documentation that enables the addition of new algorithms to each of the workflow steps.

Generation of mass spectrometry identification (MSI) standardized data format files (mzIdentML). Search engines issue

In a typical proteomic pipeline, MS data are analysed by the software to identify a set of peptides and proteins. Search engines are used routinely as part of the workflow in almost every proteomic laboratories for peptide and protein identification. Using protein sequence databases, these packages link mass spectra data to amino acid sequences and protein inference. An extensive list with the most popular and referenced tools that implements the PSI data standard mzIdentML is summarized at http://www.psidev.info/tools-implementing-mzidentml. More in detail, some commercial and free search engines offer different levels of integration with the PSI's XML-based data standardized formats.

The Mascot search engine (Matrix Science) provides full integration into the PSI's XML-based schemas. From version 2.2 (2010), it is possible to submit user's peak lists using mzML and from version 2.3 (2012), the data also can be exported using mzIdentML. This search engine may be used by most mass spectrometer platforms.

The *Sequest search engine (Thermo Fisher Scientific)* is embedded within the Proteome Discoverer software, from which it is possible to import and export using mzML format. Although there is currently no native support for mzIdentML, a stand-alone converter (ProCon) allows the export of search engine results from Proteome Discover to mzIdentML.

Other popular free search engines like Open Mass Spectrometry Search Algorithm – *OMSSA (NCBI/NLM) and X!Tandem (from GPM)*, also support mzML as input files. Open source converters have been developed independently for converting the native output formats into mzIdentML, based on the jmzIdentML Java API previously described.

Merging MS and MSI data: generation of PRIDE-XML files

PRIDE Converter 2

PRIDE Converter 2 (Cote *et al.*, 2012) (EBI) translates both MS and MSI (peptide/protein identification) data from most common data formats into valid PRIDE XML standardized format for submission to the publicly available PRIDE database. It presents a convenient, wizard-like graphical user interface, and includes efficient access to the Ontology Lookup Service (OLS) (Cote *et al.*, 2010a), required for annotating experimental results with metadata about the study. From many of the available search engine output data formats, this tool extracts and combines MS and MSI data to generate a PRIDE XML result file.

Adding metadata: generation of standardized MIAPE documents

The ProteoRed MIAPE Webtoolkit – PMWTK

Even though PRIDE Converter 2 includes user cases for many of the available search engine output data formats, it does not include all the metadata required by MIAPE guidelines. To solve this issue, PMWTK (Medina-Aunon *et al.*, 2011b) (ProteoRed-ISCIII, CNB-CSIC) offer an alternative framework where these MIAPE metadata are captured from a variety of input files and included within the generated PRIDE XML files. This information allows user to ensure both the quality and reproducibility of the data according to the MIAPE guidelines. PRIDE generation through PMWTK is done in a straightforward manner. Staring from the input data (MS and MSI) and the MS metadata files, the corresponding MIAPE MS and MSI files are created and stored within the ProteoRed MIAPE database (PMDB). MS metadata are fulfilled by the user within the main interface and it is saved for future cases. Then, from the created MIAPE document, a PRIDE XML document may be generated at the end of the pipeline.

Data analysis and reporting tools

A proteomics user-friendly framework should support data visualization. For mzML, both jmzml and jmzreader provide a simple graphical interface for viewing results. The jmzML viewer can be used to load and visualize mass spectra and chromatograms from multiple and large mzML files simultaneously. The jmzreader's interactive viewer adds the possibility to load and view spectra from all supported file formats. Both viewers illustrate the usefulness of the low-memory footprint achieved by the whole library as well as the advantage of the common interface. Besides these GUIs, other softwares are described below (Table 7.6).

Table 7.6 Data analysis tools

Tool	Standard	Link
inSilicosViewer	mzML	http://insilicos.com/products/insilicos-viewer-1
ProteoIDViewer	mzIdentML	http://code.google.com/p/mzidentml-viewer/
PRIDEViewer	PRIDE XML	http://proteo.cnb.csic.es/prideviewer/
PRIDE Inspector	PRIDE XML and mzML	http://code.google.com/p/pride-toolsuite/
OpenMS/TOPPAS	mzML and mzIdentML	http://open-ms.sourceforge.net/
Scaffold	mzIdentML	http://www.proteomesoftware.com/
MIAPE extractor	MIAPEs and PRIDE XML	http://www.proteored.org/MIAPE/

InsilicosViewer (mzML) (http://insilicos.com/products/)

The Insilicos Viewer URL (Insilicos LLC, USA) is a software tool for visualizing raw MS data, supporting various file formats such as mzML, mzXML, mzData, ANDI CDF, and Thermo (RAW) files. Insilicos Viewer runs only on Windows operating systems and it is not open source software (although it is free for users). Once the file is loaded the chromatographic information (base peak) or individual MS spectra can be displayed, offering different options such as total ion current, base peak, or an intensity map.

The ProteoIDViewer (http://www.proteosuite.org/)

In order to help the experimental proteomics community to use the mzIdentML standard, the *ProteoIDViewer* (Ghali *et al.*, 2013) was developed by the Institute of Integrative Biology at the University of Liverpool. One of the main advantages of the ProteoIDViewer is the ability to open other file formats, such as the OMSSA and X!Tandem files and automatically translate to mzIdentML. The different items reported within an mzIdentML file are depicted along the GUI's tabs, allowing users to focus on identification data from the angle of protein groups, individual spectra or peptide-spectrum matches (PSM). The tool integrates code for estimating the false discovery rate (FDR) when a decoy database search has been performed, and displays the results on graphs and global statistics tab contains summaries of the mzIdentML files, including the number of confidently identified peptides and proteins.

PRIDEViewer (http://proteo.cnb.csic.es/prideviewer/)

PRIDEViewer (Medina-Aunon *et al.*, 2011a) is a Java-based application developed (ProteoRed ISCIII-CNB/CSIC) that visualizes the information in PRIDE XML files, allowing the end user to track the proteomics pipeline from the experiment description to protein/peptide validation through an intuitive graphical interface. The presentation follows the main sections in which PRIDE itself is divided: a protein identification navigation tree, experiment description, instrument and mass spectra panels, and additional information regarding the set of identified proteins and peptides. All the displayed items are linked within the same interface preventing time-consuming steps such as browsing the public repository. In addition, PRIDEViewer brings the opportunity to validate the reported peptides and proteins by resubmitting the original mass spectrometry data to the new database version via the Mascot search engine in order to get a rapid validation of the results.

PRIDE Inspector (http://code.google.com/p/pride-toolsuite/wiki/PRIDEInspector)

PRIDE Inspector (EBI) (Wang *et al.*, 2012). This Java stand-alone GUI enables initial assessment of data quality and reliability. This web tool is a freely available open-source software tool that allows the user to efficiently browse and visualize mass spectrometry (MS) proteomics data. In addition, PRIDE Inspector supports fast loading of PRIDE XML and mzML files, and it provides direct access to all public PRIDE data through a direct MySQL database connection. Moreover, this software includes an automated data download capability for private PRIDE experiments that allows journal editors and reviewers to assess the relevant experiment(s) during peer review. PRIDE Inspector can also visualize quantification data for both protein and peptide identifications. Support for visualization of mzIdentML is currently under development.

OpenMS/TOPPAS (http://open-ms.sourceforge.net/workflow-integration/toppasworkflows/)

OpenMS (University of Tübingen) displays MS data in various views, providing a two-dimensional view for single scans, and a three-dimensional plot for LC-MS maps (retention time vs. m/z). Furthermore, built on the same platform there is another graphical aid, the TOPP Pipeline Assistant (TOPPAS) (Junker *et al.*, 2012). This tool allows a rapid composition of HPLC-MS analysis workflows, reducing it to simple drag-and-drop of analysis tools and the addition of connections in between. All TOPP tools shipped with OpenMS can be used and connected in order to build custom proteomics workflows.

Scaffold (http://www.proteomesoftware.com/products/scaffold/)

In the context of proteomics, Scaffold (Proteome Software, Inc., USA) was projected to compose a structure around the data so researchers have a flexible framework that would simplify their analyses. The goal was a software that should be intuitive and easy to use, it has been built for powerful data manipulation and uses robust, computationally-demanding statistics and automatic data recalculation based on confidence thresholds. It includes the computational infrastructure to deal with some of the most frequent lacks for proteomics users. Protein Grouping, FDR filtering, quantitative proteins or enhanced GO support are offered in a friendly and graphical environment.

Although, Scaffold is a commercial product, it offers a free viewer enables researchers to share their results with colleagues.

MIAPE Extractor (http://www.proteored.org/MIAPE/)

MIAPE Extractor is an appealing user interface where it is possible to get a huge variety of graphical charts that allow an intuitive way to inspect the experiment results. Furthermore, it allows users to adapt the data according to the experimental design. Different fractions could be defined within this framework to be individually processed (MS and MSI data). The usual pipeline includes data filtering to remove wrong identifications. Thus, if searches were done against a forward-decoy database, this tool includes methods to establish FDR threshold using different levels (PSM, Peptide or Protein). In addition, protein-grouping algorithm (Prieto *et al.*, 2012) has been included to improve the protein inference issue.

Conclusions

This chapter illustrates some of the most widely accepted initiatives to provide a better understanding of proteomics results. First, we highlighted the importance of reporting proteomics experimental data with sufficient detail to enable the replication and reanalysis of reported results. This is governed by a cohort of minimal information requirement guidelines (MIAPE), standardized data formats (XML) and semantic ontologies, all of which were proposed by the proteomics community through the HUPO-PSI working groups.

Secondly, an additional key point we addressed was the web resources available to publish and retrieve protein-related data. In line with this MS-based data generation issue, there are proteomics databases such as PRIDE, PeptideAtlas, GPMDB (now synchronized under the ProteomeXchange Consortium) and the PMDB, where either raw data and/or experimental metadata may be collected according to specific deposition rules.

Next, the different Application Programming Interfaces (APIs) are described to enhance the use of the data standards in the laboratory workflows. Third-party applications could include these APIs to either import or export normalized proteomics results, opening a new frame for data sharing within the scientific community.

References

Bertsch, A., Gropl, C., Reinert, K., and Kohlbacher, O. (2011). OpenMS and TOPP: open source software for LC-MS data analysis. Methods Mol. Biol. *696*, 353–367.

Binz, P.A., Barkovich, R., Beavis, R.C., Creasy, D., Horn, D.M., Julian, R.K. Jr., Seymour, S.L., Taylor, C.F., and Vandenbrouck, Y. (2008). Guidelines for reporting the use of mass spectrometry informatics in proteomics. Nat. Biotechnol. *26*, 862.

Brusniak, M.Y., Kwok, S.T., Christiansen, M., Campbell, D., Reiter, L., Picotti, P., Kusebauch, U., Ramos, H., Deutsch, E.W., Chen, J., *et al.* (2011). ATAQS: A computational software tool for high throughput transition optimization and validation for selected reaction monitoring mass spectrometry. BMC Bioinformatics *12*, 78.

Carr, S., Aebersold, R., Baldwin, M., Burlingame, A., Clauser, K., and Nesvizhskii, A. (2004). The need for guidelines in publication of peptide and protein identification data: Working Group on Publication Guidelines for Peptide and Protein Identification Data. Mol. Cell. Proteomics 3, 531–533.

Cote, R., Reisinger, F., Martens, L., Barsnes, H., Vizcaino, J.A., and Hermjakob, H. (2010a). The Ontology Lookup Service: bigger and better. Nucleic Acids Res. *38*, W155–160.

Cote, R.G., Griss, J., Dianes, J.A., Wang, R., Wright, J.C., van den Toorn, H.W., van Breukelen, B., Heck, A.J., Hulstaert, N., Martens, L., *et al.* (2012). The PRoteomics IDEntification (PRIDE) Converter 2 framework: an improved suite of tools to facilitate data submission to the PRIDE database and the ProteomeXchange consortium. Mol. Cell. Proteomics *11*, 1682–1689.

Cote, R.G., Reisinger, F., and Martens, L. (2010b). jmzML, an open-source Java API for mzML, the PSI standard for MS data. Proteomics *10*, 1332–1335.

Desiere, F., Deutsch, E.W., King, N.L., Nesvizhskii, A.I., Mallick, P., Eng, J., Chen, S., Eddes, J., Loevenich, S.N., and Aebersold, R. (2006). The PeptideAtlas project. Nucleic Acids Res. *34*, D655–658.

Deus, H.F., Prud'hommeaux, E., Miller, M., Zhao, J., Malone, J., Adamusiak, T., McCusker, J., Das, S., Rocca Serra, P., Fox, R., and Marshall, M.S. (2012). Translating standards into practice – one Semantic Web API for Gene Expression. J. Biomed. Inform. Aug;45(4):782–794.

Deutsch, E.W. (2010). The PeptideAtlas Project. Methods Mol. Biol. *604*, 285–296.

Deutsch, E.W., Mendoza, L., Shteynberg, D., Farrah, T., Lam, H., Tasman, N., Sun, Z., Nilsson, E., Pratt, B., Prazen, B., *et al.* (2010). A guided tour of the Trans-Proteomic Pipeline. Proteomics *10*, 1150–1159.

Deutsch, E.W., Chambers, M., Neumann, S., Levander, F., Binz, P.A., Shofstahl, J., Campbell, D.S., Mendoza, L., Ovelleiro, D., Helsens, K., *et al.* (2012). TraML – a standard format for exchange of selected reaction monitoring transition lists. Mol. Cell. Proteomics *11*, R111 015040.

Domann, P.J., Akashi, S., Barbas, C., Huang, L., Lau, W., Legido-Quigley, C., McClean, S., Neususs, C., Perrett, D., Quaglia, M., *et al.* (2010). Guidelines for reporting the use of capillary electrophoresis in proteomics. Nat. Biotechnol. *28*, 654–655.

Farrah, T., Deutsch, E.W., Kreisberg, R., Sun, Z., Campbell, D.S., Mendoza, L., Kusebauch, U., Brusniak, M.Y., Huttenhain, R., Schiess, R., *et al.* (2012). PASSEL: the PeptideAtlas SRMexperiment library. Proteomics *12*, 1170–1175.

Garwood, K., Mclaughlin, T., Garwood, C., Joens, S., Morrison, N., Taylor, C.F., Carroll, K., Evans, C., Whetton, A.D., Hart, S., *et al.* (2004). PEDRo: a database for storing, searching and disseminating experimental proteomics data. BMC Genomics *5*, 68.

Ghali, F., Krishna, R., Lukasse, P., Martinez-Bartolome, S., Reisinger, F., Hermjakob, H., Vizcaino, J.A., and Jones, A.R. (2013). Tools (Viewer, Library and Validator) that Facilitate Use of the Peptide and Protein Identification Standard Format, Termed mzIdentML. Mol. Cell. Proteomics *12*, 3026–3035.

Gibson, F., Anderson, L., Babnigg, G., Baker, M., Berth, M., Binz, P.A., Borthwick, A., Cash, P., Day, B.W., Friedman, D.B., *et al.* (2008). Guidelines for reporting the use of gel electrophoresis in proteomics. Nat. Biotechnol. *26*, 863–864.

Gibson, F., Hoogland, C., Martinez-Bartolome, S., Medina-Aunon, J.A., Albar, J.P., Babnigg, G., Wipat, A., Hermjakob, H., Almeida, J.S., Stanislaus, R., *et al.* (2010). The gel electrophoresis markup language (GelML) from the Proteomics Standards Initiative. Proteomics *10*, 3073–3081.

Griss, J., Reisinger, F., Hermjakob, H., and Vizcaino, J.A. (2012). jmzReader: A Java parser library to process and visualize multiple text and XML-based mass spectrometry data formats. Proteomics *12*, 795–798.

Helsens, K., Brusniak, M.Y., Deutsch, E., Moritz, R.L., and Martens, L. (2011). jTraML: an open source Java API for TraML, the PSI standard for sharing SRM transitions. J. Proteome Res. *10*, 5260–5263.

Hermjakob, H., Montecchi-Palazzi, L., Bader, G., Wojcik, J., Salwinski, L., Ceol, A., Moore, S., Orchard, S., Sarkans, U., von Mering, C., *et al.* (2004). The HUPO PSI's molecular interaction format – a community standard for the representation of protein interaction data. Nat. Biotechnol. *22*, 177–183.

Hoogland, C., O'Gorman, M., Bogard, P., Gibson, F., Berth, M., Cockell, S.J., Ekefjard, A., Forsstrom-Olsson, O., Kapferer, A., Nilsson, M., *et al.* (2010). Guidelines for reporting the use of gel image informatics in proteomics. Nat. Biotechnol,. *28*, 655–656.

Jones, A.R., Carroll, K., Knight, D., MacLellan, K., Domann, P.J., Legido-Quigley, C., Huang, L., Smallshaw, L., Mirzaei, H., Shofstahl, J., and Paton, N.W. (2010). Guidelines for reporting the use of column chromatography in proteomics. Nat. Biotechnol. *28*, 654.

Jones, A.R., Eisenacher, M., Mayer, G., Kohlbacher, O., Siepen, J., Hubbard, S., Selley, J., Searle, B., Shofstahl, J., Seymour, S., *et al.* (2012). The mzIdentML data standard for mass spectrometry-based proteomics results. Mol. Cell. Proteomics

Junker, J., Bielow, C., Bertsch, A., Sturm, M., Reinert, K., and Kohlbacher, O. (2012). TOPPAS: A Graphical Workflow Editor for the Analysis of High-Throughput Proteomics Data. J. Proteome Res. *11*, 3914–3920.

Kaiser, J. (2002). Proteomics. Public–private group maps out initiatives. Science *296*, 827.

Kessner, D., Chambers, M., Burke, R., Agus, D., and Mallick, P. (2008). ProteoWizard: open source software for rapid proteomics tools development. Bioinformatics *24*, 2534–2536.

Kohlbacher, O., Reinert, K., Gropl, C., Lange, E., Pfeifer, N., Schulz-Trieglaff, O., and Sturm, M. (2007). TOPP – the OpenMS proteomics pipeline. Bioinformatics *23*, e191–197.

Martens, L., Hermjakob, H., Jones, P., Adamski, M., Taylor, C., States, D., Gevaert, K., Vandekerckhove, J., and Apweiler, R. (2005). PRIDE: the proteomics identifications database. Proteomics *5*, 3537–3545. Erratum in: Proteomics *5*, 4046.

Martens, L., Chambers, M., Sturm, M., Kessner, D., Levander, F., Shofstahl, J., Tang, W.H., Rompp, A., Neumann, S., Pizarro, A.D., *et al.* (2011). mzML – a community standard for mass spectrometry data. Mol. Cell. Proteomics *10*, R110 000133.

Martinez-Bartolome, S., Medina-Aunon, J.A., Jones, A.R., and Albar, J.P. (2010). Semi-automatic tool to describe, store and compare proteomics experiments based on MIAPE compliant reports. Proteomics *10*, 1256–1260.

Martinez-Bartolome, S., Deutsch, E.W., Binz, P.A., Jones, A.R., Eisenacher, M., Mayer, G., Campos, A., Canals, F., Bech-Serra, J.J., Carrascal, M., *et al.* (2013). Guidelines for reporting quantitative mass spectrometry based experiments in proteomics. J. Proteomics

Medina-Aunon, J.A., Carazo, J.M., and Albar, J.P. (2011a). PRIDEViewer: a novel user-friendly interface to visualize PRIDE XML files. Proteomics *11*, 334–347.

Medina-Aunon, J.A., Martinez-Bartolome, S., Lopez-Garcia, M.A., Salazar, E., Navajas, R., Jones, A.R., Paradela, A., and Albar, J.P. (2011b). The ProteoRed MIAPE web toolkit: a user-friendly framework to connect and share proteomics standards. Mol. Cell. Proteomics *10*, M111 008334.

Orchard, S., and Ping, P. (2009). HUPO World Congress Publication Committee meeting. August 2008, Amsterdam, The Netherlands. Proteomics *9*, 502–503.

Orchard, S., Hermjakob, H., and Apweiler, R. (2003). The proteomics standards initiative. Proteomics *3*, 1374–1376.

Orchard, S., Binz, P.A., and Hermjakob, H. (2009). Second Joint HUPO publication and Proteomics Standards Initiative workshop. Proteomics *9*, 4426–4428.

Orchard, S., Binz, P.A., Borchers, C., Gilson, M.K., Jones, A.R., Nicola, G., Vizcaino, J.A., Deutsch, E.W., and Hermjakob, H. (2012). Ten years of standardizing proteomic data: a report on the HUPO-PSI Spring Workshop: April 12–14th, 2012, San Diego, USA. Proteomics *12*, 2767–2772.

Paik, Y.K., Jeong, S.K., Omenn, G.S., Uhlen, M., Hanash, S., Cho, S.Y., Lee, H.J., Na, K., Choi, E.Y., Yan, F., *et al.* (2012). The Chromosome-Centric Human Proteome Project for cataloging proteins encoded in the genome. Nat Biotechnol *30*, 221–223.

Pedrioli, P.G., Eng, J.K., Hubley, R., Vogelzang, M., Deutsch, E.W., Raught, B., Pratt, B., Nilsson, E., Angeletti, R.H., Apweiler, R., *et al.* (2004). A common open representation of mass spectrometry data and its application to proteomics research. Nat. Biotechnol. *22*, 1459–1466.

Prieto, G., Aloria, K., Osinalde, N., Fullaondo, A., Arizmendi, J.M., and Matthiesen, R. (2012). PAnalyzer: a software tool for protein inference in shotgun proteomics. BMC Bioinformatics *13*, 288.

Reisinger, F., Krishna, R., Ghali, F., Rios, D., Hermjakob, H., Antonio Vizcaino, J., and Jones, A.R. (2012). jmzIdentML API: A Java interface to the mzIdentML standard for peptide and protein identification data. Proteomics *12*, 790–794.

Segura, V., Medina-Aunon, J.A., Guruceaga, E., Gharbi, S.I., Gonzalez-Tejedo, C., Sanchez del Pino, M.M., Canals, F., Fuentes, M., Casal, J.I., Martinez-Bartolome, S., *et al.* (2013a). Spanish human proteome project: dissection of chromosome 16. J. Proteome Res. *12*, 112–122.

Segura, V., Medina-Aunon, J.A., Mora, M.I., Martinez-Bartolome, S., Abian, J., Aloria, K., Antunez, O., Arizmendi, J.M., Azkargorta, M., Barcelo-Batllori, S., *et al.* (2013b). Surfing Transcriptomic Landscapes. A Step beyond the Annotation of Chromosome 16 Proteome. J. Proteome Res.

Taylor, C.F., Paton, N.W., Garwood, K.L., Kirby, P.D., Stead, D.A., Yin, Z., Deutsch, E.W., Selway, L., Walker, J., Riba-Garcia, I., *et al.* (2003). A systematic approach to modeling, capturing, and disseminating proteomics experimental data. Nat. Biotechnol. *21*, 247–254.

Taylor, C.F., Paton, N.W., Lilley, K.S., Binz, P.A., Julian, R.K., Jr., Jones, A.R., Zhu, W., Apweiler, R., Aebersold, R., Deutsch, E.W., *et al.* (2007). The minimum information about a proteomics experiment (MIAPE). Nat. Biotechnol. *25*, 887–893.

Taylor, C.F., Binz, P.A., Aebersold, R., Affolter, M., Barkovich, R., Deutsch, E.W., Horn, D.M., Huhmer, A., Kussmann, M., Lilley, K., *et al.* (2008). Guidelines for reporting the use of mass spectrometry in proteomics. Nat. Biotechnol. *26*, 860–861.

Vizcaino, J.A., Cote, R.G., Csordas, A., Dianes, J.A., Fabregat, A., Foster, J.M., Griss, J., Alpi, E., Birim, M., Contell, J., *et al.* (2013). The PRoteomics IDEntifications (PRIDE) database and associated tools: status in 2013. Nucleic Acids Res. *41*, D1063–1069.

Walzer, M., Qi, D., Mayer, G., Uszkoreit, J., Eisenacher, M., Sachsenberg, T., Gonzalez-Galarza, F.F., Fan, J., Bessant, C., Deutsch, E.W., *et al.* (2013). The mzQuantML data standard for mass spectrometry-based quantitative studies in proteomics. Mol. Cell. Proteomics

Wang, R., Fabregat, A., Rios, D., Ovelleiro, D., Foster, J.M., Cote, R.G., Griss, J., Csordas, A., Perez-Riverol, Y., Reisinger, F., *et al.* (2012). PRIDE Inspector: a tool to visualize and validate MS proteomics data. Nat. Biotechnol. *30*, 135–137.

Clinical Protein Science and Targeted Mass Spectrometric Assays – New Frontiers in Disease Link and Biobanking

8

Ákos Végvári and György Marko-Varga

Abstract

Today's societies have to reach out for improved healthcare solutions due to the increased number of patients and changed disease profiles represented by the dominance of cardiovascular and neurodegenerative diseases as well as various forms of cancer that require customized therapies. There is an extensive need for diagnostic and therapeutic biomarkers that may also become targets for drug developments in the future. Modern analytical platforms headed by mass spectrometry offer enhanced sensitivity as well as novel methodologies with extreme specificity to identify and quantify these potential protein markers in clinical studies, which is in the focus of international research organized by the Human Proteome Organization. However, the complexity of clinical samples and the wide concentration range of proteins represent analytical tasks yet need to be accomplished. In this chapter we present the challenges the healthcare systems face nowadays and outline possible solutions using biobanking standardization, personalized medicine and mass spectrometric technologies with relevant examples.

The unmet needs and progresses in modern healthcare

Today's society has increased its demands and expectations on healthcare quality and efficiency. This happens during a period of fast technology progress and developments with great advancements in modern life sciences. These changes in time currently open up for new opportunities that can meet challenges to the medical research community. The aim and objective is to develop and progress a strategy with patient-centric and technology-driven research.

Diagnostic biomarkers are examples of a biomedical priority area that is expected to deliver and provide accurate indices of health status to be used to assist clinical decision-making. Currently clinical doctors are challenged with complex disease presentations of patients and the diagnosis needs to be based on poor and inadequate analytical data generated in central hospital laboratories.

The dimensions of diagnostic biomarkers in the development of accurate and reproducible quantitative indications further require measures that can be compared in large patient cohorts. Here, standardization is a key element that is needed to provide assurance of the quality of the clinical samples providing these biomarkers to form the basis for an accurate patient diagnosis. A main objective at this stage is also to insure that the technology platforms used to provide such biomarker measurements also are consistent in the diagnostic read-outs. The measurement being performed on an instrumental platform in India should provide the same level of accuracy as an instrument in Sweden, England or the United States. This global quality assurance needs to be established, maintained and progressed to make global healthcare successful.

Clinical samples that have been collected for clinical analysis can either be used immediately, or discarded. Lately the build of large-scale biobanks is being built for future discovery research and clinical use (Marko-Varga, 2013; Marko-Varga *et al.*, 2012; Végvári *et al.*, 2011, 2012). The current methods of storage of samples can differ significantly in different laboratories, hospital institutions and government agencies. Consequently, one of the challenges encountered in large-scale biobanking activities has been the difficulty in inter-laboratory studies. Differences in, for instance experimental designs, protocols, as well as reagents and disposables used will clearly impact on data the quality and reproducibility generated in clinical evaluations. Standardization and best practice protocols has been proposed to address this problem. Consequently, the strategic roles in biobanking are a matter at heart where major investments are being made, whereas new directives and best practice with clear standardizations and procedures needs to be applied.

Today's general practice in hospitals relates to a situation where clinical samples are managed in biobanks spread throughout the world for biomedical research, in both the public and private sectors. A general perception in modern healthcare is that each biobank follows its own protocols and in general implements its own quality criteria. This subsequently leads to a large variation on the way that the quality, integrity and performance of individual samples can vary extensively, and becomes a major challenge. A significance of the current status is that experimental studies using these resources therefore lose statistical power and significance. An evaluation of the current tools at hand in biobanking and the resulting recommendations for improving this status quo is a very important first step in providing a best-case analysis that can be used as a model for improving current clinical practice that will benefit the patients. It is likely that such evaluations will point out local commendation linked to those local circumstances. Irrespective whether the initiatives within biobanking is being conducted on a small scale or on a large scale, the area of best practice in biobanking has many common elements that are universal to all of these archiving activities. We do foresee that in the future, even the physical structure of sample storage will have the most common standard and integration with other areas of patient care integrated where the healthcare status is linked to the overall patient treatments. In this strategy, it becomes the development of consensus of procedures and policy. These building components are made in order to guide future questions of the how, where and why samples will be utilized, providing benefit to the patient. These timely developments were highlighted by *Time* magazine that published an article on the 'Ten Ideas Changing the World Right Now'. In this publication, biobanking was highlighted as an important activity with important potential consequence for mankind (Park, 2009).

The challenge at hand is to have the potential to build large standardized and highquality patient sample collections that offer information that can contribute as a critical factor for successfully developing new drug and diagnostics developments. Biobanks will fill in an important gap here that is strategic and key for future healthcare developments.

The progress of new problem-solving biomarkers have a great promise, where both the industry and the academic fields are investing and exploring approaches to exploit technology to make innovative discoveries.

Lately, the protein science research field, also referred to as the proteomics area, has developed new and complementary technologies, like protein shotgun sequencing and quantitative mass spectrometry platforms that are readily available. It is being utilized in everyday clinical chemistry assays within hospitals around the world. With about 20,300 gene products and their multiple structural variations involved in disease pathophysiology, the task to measure these proteins is a real challenge to the research society. The coding of the human genome is with its splice variants expected to produce a much higher number of human proteins. The exact human protein number is still not known today. However, considering the post-translational modifications that occur in the body, it is expected to be many millions of proteins. The cellular activity on all these proteins in any given disease state, forms a high degree of complexity that holds the holy grail of many diseases. The heart of the matters is that proteins are, unlike the human genes, dynamic targets that constantly change not only their relative abundance levels but also their physical forms. This is one important reason why the protein area has a much higher complexity and more variable in human populations. In this respect, the resting steady state of a protein may change its form and function during a disease development such that the activation state of a protein is perturbed by in most situations the post-translational modifications of the protein sequence by, for example, phosphorylation, glycosylation, oxidation, alkylation and acylation.

Since the expression of proteins in human biofluid is a rich source of clinical material for patient analysis, healthcare has given blood analysis and protein quantitation a high degree of attention over the last decades. Blood samples like plasma and serum constitute a protein expression range of 10–12 orders of magnitude. This is a real challenge, since there is no analytical technology platform today that can measure such a large dynamic range of any given protein in patient samples (Anderson and Anderson, 2002; Domon and Aebersold, 2006). The limitation of sensitivity of currently available protein platforms does not allow us to perform such a broad assay. The alternative solution that most laboratories apply is to build the assays in a way that covers 3–5 orders of magnitude in protein concentration. Multiple concentration ranges can be developed that each separately covers the different ranges that finally covers a broad 8–10 order of magnitude read-out.

Right now there are a large numbers of recognized markers that are used for various diagnostic predictions. The goal here is to assess which combination of markers has the greatest diagnostic and prognostic power, and that can aid the patient. As there is an ever-increasing effort in the clinical field to try to understand and predict the disease status and malfunctions of patients, the focus on optimal patient drug treatment has become a central point of attention. The novelty of personalized medicine with targeted mode-of-drug-action is also a key development in modern healthcare that is driving the need for new diagnostic biomarkers and large patient sample cohorts that are being archived in biobanks.

There is an increasing international interest to strengthen and progress research areas that can aid in solving disease understanding and to improve patient care. The 'one drug for all' concept does not work as well as before, as the demand from modern society has increased. Instead, new approaches are needed such as tailored medication summarized within all the parts that are related to personalized medicine. This new, targeted drug treatment has opened up an entire new field, where the patients within a given disease are considered as a heterogeneous rather than one uniform group that can be treated in the same way. Instead, an alternative dedicated solution is targeted for a specific phenotype. This new generation of medications is at the frontline of modern medicine and most drug companies have the pipeline activated in clinical trials to evaluate and adapt these treatments to new patient groups. Personalized medicine is closely linked to the diagnostic arm of the treatment, when the selection is made. Proteomics and the clinical protein science fields are devoting lots of resources to outline the optimal path, linking the treatment and diagnosis together combining the novel treatment technologies with an early indication of disease diagnosis utilizing both imaging techniques and biomarker diagnostics (Kato *et al.*, 2011).

Although considerable progresses are being made in modern medical research, there is still an unmet need for solutions, how best to manage and address these healthcare challenges. These actions are well known to governments and health care institutions. These are reasons why for instance the European Commission (EC) has dedicated large-scale research programs to address the developments of disease mechanism research within dedicated research area. In addition, the National Institutes of Health (NIH) in USA, as well as other sponsoring bodies in the world, is adding into these resource builds. In a joint effort between Europe and the US, common strategies of how systems biology can be useful in cancer research have been developed.

Recently, manifestations on the expectations of future healthcare has been presented, a lot of promises has been announced, including Wall Street expectations on biomarkers have yet to be manifested (Landers, 2003). The technology driven disease biology proved to be a greater challenge than expected. Another great endeavour has been started and initiated: the Human Proteome Project (HPP) that was launched in September 2010 in Sydney at the HUPO World Congress (Hancock *et al.*, 2010). This idea and science project outline was already presented by Anderson and Anderson several decades ago (Anderson *et al.*, 2001). The current status of the Chromosome-based HPP (C-HPP) is that it has initiated a consolidation of protein-knowledge, to build an integrated, structured encyclopaedia of human proteins under international collaboration of a global C-HPP Consortium comprising 22 countries throughout the 24 chromosomes. Recently, the C-HPP organization made a joint engagement with the Encyclopedia of DNA Elements, the 'ENCODE' consortium, in order to integrate the genomic coding of proteins and elucidate the protein functions (Paik and Hancock, 2012).

Building new medical frontiers in modern society

We are experiencing that the social welfare systems, which are delivering the daily medical care, are today in a state of major restructuring and change. In order to meet the demand in everyday healthcare, we realize that we are lacking both resources, as well as targeted treatment efficiencies, that provide an improved quality of life for patients. These alterations can be noticed by the fact that many high-quality treatments in most common diseases, such as cancer, cardiovascular diseases, neurodegenerative diseases and diabetes, to mention the most resource demanding, are not being cost-effective. The number of patients around the world is increasing due to the rapid growth of the elderly generation (65+ age group). This is certainly a global trend and development, rather than local needs (Kato *et al.*, 2011). Considering biobank resources as an added value to build the future healthcare, some positioning in society and clarification requirements arises.

Biobanking is not a new idea or concept. In fact, blood sample collections have been an integral part of medical care for a long time with long standing traditions. The experience of sampling and storing whole blood and blood products has made great advancements not the least of which are the registers of healthy volunteers that provide the samples and maintain the value. In Scandinavia, medical doctors in hospitals have also been collecting samples for more than a hundred years. The ultimate goal of these initiatives has been to get a better understanding of the presentation of disease within patient groups and how best to understand the correlation to clinical measurements.

Eventually, in relation to these healthcare visions and concepts, it is expected that each society is able to offer improved prognosis, at a reduced cost to the healthcare system by early disease indication, with personalized treatment and evaluation of responses to treatments.

Impact to society of personalized medicines

Targeted treatments are becoming the new standard for disease treatments where each disease is divided into patient responder groups. The new generation of drugs is specific in a sense that they provide improved patient efficacy and better patient safety, especially within cancer, such as lung, breast, as well as malignant melanoma cancers. In these cancers, the success of personalized medicine has shown a considerable increase in quality of life for patients during the last decade.

4P Medicine is a novel concept that is gaining more interest as well as progress, where basic research is being combined with clinical treatment. The 4P Medicine encompasses predictive, preventative, personalized and participatory features in the future build and improvement of treatment (Fitzgerald *et al.*, 2006; Weston and Hood, 2004).

Within pulmonary oncology, the observation of EGFR somatic mutations in lung cancer patients were made at a time point during the phase III study with IRESSA for the first time in North America in 2005. The phase III study was not able to show a statistical significance of the drug with the criteria that were set at the time. However, previously, in the phase II study, the study outcome was highly successful with the dose and conditions given at the time, while the drug company, AstraZeneca was urged to start production and distribution to patients before the phase III study was finalized. Nonetheless, at this stage, the subset of the Asian part of the patient cohort did prove a significant outcome upon IRESSA treatment, while the mode-of-drug-action already at this time point was suspected to be related to EGFR-mutation. The test to analyse the EGFR mutation status was developed that could identify one phenotype: mutated or non-mutated (Lynch *et al.*, 2004). These findings were associated with a long path of mechanistic development works. The resulting data from these studies and additional follow-up investigations resulted in a standard procedure using the EGFR-mutation assay, which could detect mutations in exon18 (G719A/C/S), exon19 (E746-A750 deletion, L747-P753 deletion insertion S), exon20 (S768I) and exon21 (L858I, L861Q) more than 1% mutation rate in the clinic today. In Japan, a positive indication will result in personalized medicine prescriptions (Kosaka *et al.*, 2004; Lynch *et al.*, 2004; Paez *et al.*, 2004a).

In protein expression research today, we discover, develop and validate proteomics findings for new prospect applications in a clinical setting (Marko-Varga *et al.*, 2007; Nyberg *et al.*, 2011). Statistical analyses of the utility of protein biomarkers in predicting acute interstitial lung disease (ILD) in drug-treated patients from an epidemiological case–control study (Nyberg *et al.*, 2011). As there are increasing funding globally available for research programs that can improve the diagnosis and stratification of patients, as well as biomarkers for both safety assessments and efficacy, an increasing number of studies and study data that illustrate these developments.

Gefitinib (IRESSA®, ZD1839) was developed as a specific inhibitor of the EGFR tyrosine kinase. Gefitinib target the EGFR for therapeutic drug intervention within lung cancer. Early reports appeared on the experience of lung cancer patients with EGFR mutations. These were among the first clinical data that provided correlation with clinical response to gefitinib therapy (Paez *et al.*, 2004b). Personalized medicines have an improved efficacy over chemotherapy and radiation treatments. While targeted drugs can reach 70–75% efficacy, the combined effects of lung cancer chemotherapy and radiation, will reach about 35%. With respect to the well-being of patients, the targeted drugs are not only more efficient, but also provide an improvement in quality of life.

In addition various prognostic factors of gefitinib have been concealed, such as the Asian, EGFR somatic mutation, non-smoking and woman (Kudoh *et al.*, 2008). Recently a clinical study in four Asian countries was conducted: the IRESSA Pan-Asia Study, the 'IPASS' study, in which 1217 NSCLC patients were enrolled (Mok *et al.*, 2009). The IPASS study was an open label, randomized, parallel-group study that assessed the efficacy, safety and tolerability of IRESSA versus doublet chemotherapy (carboplatin/paclitaxel) as first line treatment in a clinically selected population of patients from Asia.

The adenocarcinoma cohort of the study with Japanese patients was enrolled for efficacy biomarker discovery. The ultimate objective is to identify responders to gefitinib treatment to non-responders. The final outcome for diagnostic prediction is still a challenge and under investigation.

Biobanking

Sample storage in healthcare institutions

Biobanking has nowadays been identified as a key area for development in order to accelerate the discovery and development of new drugs. It is clear from public legislation and investment the establishment of biobanks around the world has become an integrated part of modern healthcare (Betsou *et al.*, 2010).

In most countries biobanking is organized as a core facility within the hospital clinical chemistry structure, rather than a segregated activity as a stand-alone operation. Biobank activities are closely associated with pathology and diagnostic activities. In some countries we can see that the Biobank has become an autonomous part of the healthcare industry. The biobank concept is in a phase of development where the implementation into the clinical organization is ongoing, with a varying degree of integration in Europe, North America and Asia.

In Europe, the Biobanking and Biomolecular Resources Research Infrastructure (BBMRI), a joint initiative which involves more that 200 organizations in 24 EU member states, is planning an EU infrastructure (http://www.bbmri.eu). The aim and objective here for the BBMRI consortia as a EU resource is to build sustainable and accessible biological resources that is required for health-related research and development intended to improve the prevention, diagnosis and treatment of disease and to promote the health of the citizens of Europe.

In Scandinavia, but particularly in Sweden and Denmark, there has been a long tradition of longitudinal epidemiological studies within the general population. The Swedish Twin Registry, twhich was started in 1960, is by far the largest registry in the world with currently over 86,000 twin pairs under current study (Lichtenstein *et al.*, 2002; Svartengren *et al.*, 2009). Along with the sample collections in major biobanks, clinical data and extensive archive information from the participants in the study are collected in large national registers. Another Swedish national resource is the cancer registry, all of them that were collected cases of cancer, going back since 1958. All of these data and sample collections can then be used by relating them to the cause of death registry, and all underlying causes This is an invaluable and important asset that is being built into modern healthcare systems. Nowadays it is also possible to follow and provide data that relates to the medical history of patients along with the medical Birth Registry. These extents of these medical resources are probably in the absolute frontline of international standards, and form a model for international followers. The ability to align large data registers with everyday treatments of patients is absolutely necessary and is expected to grow considerably in the near future. The benefit to patients will be the utility to align biobank sample output, to pathological findings and correlations that can aid in modern disease treatments, where the patients will be the great benefiters. Still, an estimated $1 billion has been invested in the biobanking industry within the last ten years. At least 179 biobanks with 345,000 donors exist in the USA, most of which were established in the last ten years (source: Business Insights, March 2009).

Biobanks are a major resource for scientists to access unique patient samples for medical research. Many studies within clinical proteomics utilize stored samples contained in biobanks to measure specific end points. Today we have not yet achieved consensus on how to collect, manage, and build biobank archives in order to reach goals where these efforts are translated into value for the patient. Standardizations of quality control of samples being processed for storage as well as retrieval of stored samples are important goals in order to support the development of diagnostic tools such as biomarkers. Lack of available high-quality sample collections with a wide range of disease cohorts has become a rate-limiting step for drug development, medical research and novel diagnostics. Various technology platforms are expected to play an important role in ensuring the need to preserve the sample integrity of biosamples stored for periods that reach one or several decades. The usefulness of samples in comparative studies will also be heavily influenced by the levels of annotation describing clinical status and metadata of measurements of clinical phenotype. Recently several international initiatives have been launched to disseminate best practice examples for biobanking. These developments will be of great value and importance to programmes such as the C-HPP that will associate protein expression in healthy and disease states with genetic foci along of each of the human chromosomes.

Large-scale international biobanks

Millions of clinical samples are obtained every day for use in diagnostic tests that support clinical decision-making. Clinical samples can also be archived into repositories for use in future studies investigating the root causes of disease using genetic, genomic, proteomic and metabolomic approaches. Today, it is estimated that over one billion clinical samples are assembled into so-called biobanks, also known as biospecimen resources, and stored throughout the world (Lasso, 2010). Implied but not visible within these sample number estimates are the costs in resources (material, financial, and human) required to establish and maintain biobanks. In 2009, *Time* Magazine highlighted the importance of biobanks as '10 Ideas Changing the World Right Now'. Two years later, we are still developing our understanding as how to best utilize the existing resources and how to develop them further. Biospecimen resources are found in many forms and serve many purposes. In rough terms, we can divide biobanks into two categories:

1 samples collected before year 2000 that represent for the most, manually processed samples that have been stored frozen at −20°C or −80°C in freezers located within general laboratories, or as tissue block collections in pathology labs; or
2 samples collected in the last decade and processed using automated sample handling followed by cold storage in dedicated facilities at deep freezing temperatures (−80°C and −260°C).

Most biobanks are funded publicly and sample storage often becomes a question of budget resources. The biobanks could represent a small-scale collection of a few samples of selected observation points made by an individual investigator. Or the collections may be part of large studies sponsored by an academic hospital or part of a multicentre national study that assembles and shares samples within consortiums of investigators studying different aspects of a disease process. Commercial biobanks are also growing in number, especially in areas where pharmaceutical development is focused. Publicly funded support to develop the infrastructure of large bio-specimen collections held at multiple sites in multi-national studies is also becoming more common today. Some of these repositories are solidly emplaced in our systems of normal care such as blood banks, organ banks used in transplantation, and sperm and ova used in fertilization. Beyond their use in clinical care, these collections also support research activities, such as the investigation of disease mechanisms in both rare and common diseases, measurements of biomarkers operating during complex diseases, measurement of markers for risk of disease in population based surveillance studies and in clinical trials monitoring the effects of drugs on eventual outcome.

This latter activity within clinical trials has gained much attention as new treatment paradigms such as personalized medicine are put into standard practice. As recently argued by leaders of the National Institutes of Health (NIH) and the Food and Drug Administration (FDA), 'The success of personalized medicine depends on having accurate diagnostic tests that identify patients who can benefit from targeted therapies'. The search for biomarkers for use in personalized medicine therapies is a highly active area, bridging disease areas, clinical specialties and drug development but is highly dependent upon gaining access to clinical samples of high quality and in statistically relevant numbers for study comparisons.

Large-scale biobank developments have been established within the Biomedical Center in Lund, Sweden, with biofluid archiving of patients. Blood fractions from patients, comprising whole blood, serum, plasma (EDTA, citrate, and heparin), are being stored and processed by automated processing in −80°C storage units. High-density sample tube systems are applied to maximize the sample output per patient utilizing 384-rack systems (Végvári *et al.*, 2012). We have built into the process, barcode assignments that ensure the uniqueness of each and every sample that is used within the analysis process generating both MS-MS and antibody read-outs. The large-scale biobanking workflow is depicted in Fig. 8.1, showing the processing of biobank samples made in order to generate extensive protein sequence information in health and disease. These difficult healthcare issues will entail large

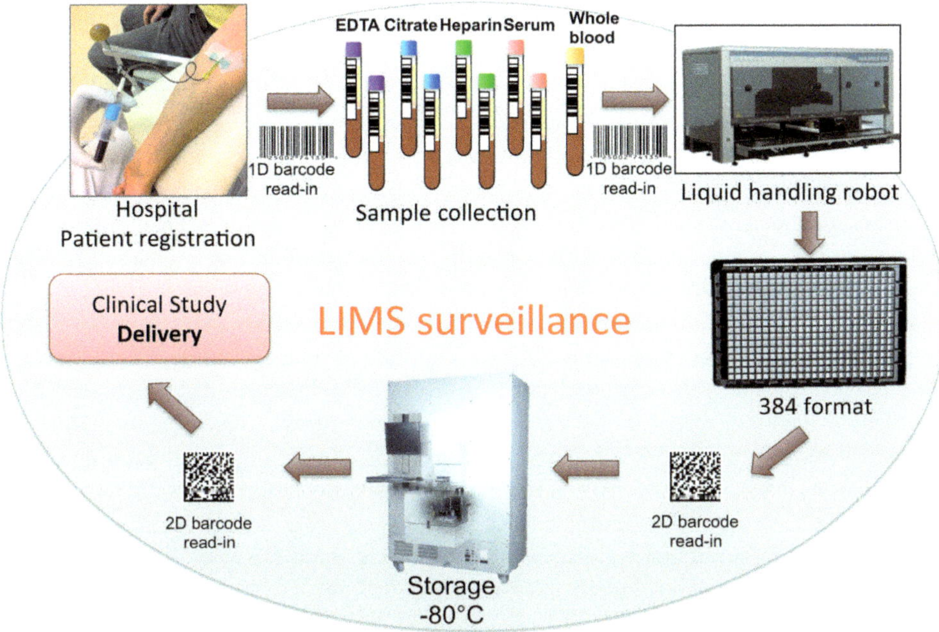

Figure 8.1 Illustration of the biobanking workflow.

number of biobank samples with ample statistical power. Worldwide biobank archives will need to work together and exploit these cohort sample assets in order to generate value in future science milestones that will transform value to the patient. Biobanks of human patient samples tissues and blood fractions are increasingly recognized as major assets in disease research. These new sample archives are built within both academic, commercial health care, the drug development industry, as well as within the government regulating agencies.

Biobank samples are increasingly used as a major resource for scientists to access unique patient samples for medical research and patient treatments. Many studies within clinical proteomics utilize stored samples contained in biobanks to measure specific protein expressions. For Biobank samples stored in either −20°C or −80°C (the preferable temperature), there is still today a debate on how to reach a consensus on how best to provide:

- management and storage conditions;
- sampling procedures and protocols;
- specific sample and data assignments to sample tubes linked to the barcode identifier;
- laboratory intelligence management systems and IT/IS solutions allowing sample tracking.

All of these efforts are made in order to reach goals where these efforts are turned into value for the patient. Standardizations as well as quality control of samples being processed for storage are key parts that need to be established and introduced in modern biobanking. The retrieval of stored samples are the next steps of important goals that needs to be put in place in order to support the development of both drug targets as well as diagnostic tools such as biomarkers.

Todays major challenge and the highest unmet need in modern healthcare is the monitoring and prediction of response to treatment. The costs associated here for the society is huge. The optimization of drug treatment that easily can be adjusted for by scheduling the dosing, discontinuing, the drug treatment, and thereby optimize and improve on the safety and/or efficacy of a targeted drug treatment. Consequently, the status today is that the lack of available high quality sample collections with a wide range of disease cohorts currently is a rate-limiting step for drug development, medical research and novel diagnostics.

Ultimately, the C-HPP global project initiative will aid by increasing the awareness within the proteomics community regarding the basic concepts of clinical samples. Understanding and standardizing the qualitative requirements of clinical samples in order to generate protein analysis data that can be globally shared and used for future developments within our society will also form the basis for improving on the data quality of proteomic studies. This means that the drug development processes that include drug design and synthesis followed by physicochemical property optimization of drugs, is currently the strong part of pharma pipeline strategy. Consequently, it is imperative to understand the role of drugs and what they do in patients. Areas such as:

- the link of mode of drug action to the pathophysiology;
- the disease link;
- disease mechanisms that relates to molecular pathways;
- genetics and selected Protein targets with high drug affinities.

That needs to be aligned with the:

- patient stratification selecting the right phenotype(s);
- potential responders within the patient cohort;
- individual risk assessments.

The targeted personalized medicine developments ultimately will need to align with the increasing safety demands that are key for preventative medicine and early treatment. Another challenge for new drug developments, in which the C-HPP deliveries can play a role, are disease phases and complex pathophysiology that needs to be understood for any given treatment.

Protein biomarkers and targeted analysis

Protein expression analysis

Studies are conducted by global expression analysis, where a non-directed principle is applied, where many thousands of proteins and/or small molecules can be analysed and sequenced in a small amount of sample. Studies where the analytes of interest are known, is measured by a targeted approach, where a specific and smaller set of analytes are measured in dedicated assays. In the last years, multiple reaction monitoring (MRM) assays have become very popular due to their generic concept (Huttenhain *et al.*, 2009; Lange *et al.*, 2008; Picotti *et al.*, 2010; Surinova *et al.*, 2011).

Patient needs and disease symptoms are very often multifactorial. Consequently, the demands are driven by more than one biological mechanism. In addition, disease presentations are in most cases not a one-point position of a treatment event. On the contrary, it is a situation where diverse variable disease presentation by the complex pathophysiology. These are reasons why the diagnosis of the disease is a highly important part of the health care process. The diagnosis as such, appears in the very beginning when the patients first meet the doctor. A number of analysis and investigations are performed, most commonly involving clinical chemistry assay measurements. In many cases, imaging tests are made, such as X-ray, CT, PET and MRI. The responsible physician will have consultation throughout the disease progression and treatments, where the diagnostic tests following the progress of the disease will be repeated several times throughout the treatment of the disease.

Multiple reaction monitoring (MRM) assays

Biobanking developments provide large amount of clinical samples for analysis of protein biomarkers, which are recognized as differentially expressed in comparing clinical status of disease and health. Currently, mass spectrometry (MS) is the most frequently applied protein sequencing and detection platform when interfaced to liquid chromatography (LC). Quantitation can be performed by both labelled as well as non-labelled LC-MS profiling technologies when applied to protein, peptide, and metabolite profiling and differential expression analysis. Selected reaction monitoring (SRM) mass spectrometry is a sequencing technology platform, where in most cases, triple quadrupole tandem mass spectrometry principles are utilized for the quantitation of proteins and peptides assays applied within the clinical field (Huttenhain *et al.*, 2009; Nishimura *et al.*, 2010; Picotti *et al.*, 2010). This mass spectrometry technology platform allows for absolute protein quantification in very minute amounts of bio-fluid samples with multiplexed read-outs of disease panels (Rodriguez *et al.*, 2010).

Following biomarker validations, SRM offers quantifications of proteins in complex biological matrices measuring peptide levels. In combination with appropriate stable isotope-labelled internal standards, the SRM approach provides absolute quantitation of the analyte. Additionally, a high number of proteins of interest can be monitored simultaneously in SRM assays (Anderson, 2005).

Prostate cancer protein biomarker analysis

The SRM quantifications present high sensitivity and speed, which is a future requirement for high throughput screening of clinical samples for candidate biomarkers. Currently, SRM applications are the fastest growing targeted protein analysis area, with multiplex assays for absolute quantitation in clinical disease areas. For these reasons, we utilize the SRM technology in quantitation of prostate-specific antigen (PSA) isoforms in clinical samples.

PSA is recognized as useful biomarker of prostate cancer and has been repeatedly studied in SRM method development as model protein (Addona *et al.*, 2009; Barnidge *et al.*, 2004; Enjalbert *et al.*, 2013; Fortin *et al.*, 2009a,b; Keshishian *et al.*, 2007; Kulasingam *et al.*, 2008; Li *et al.*, 2011; Liu *et al.*, 2012; Paik *et al.*, 2012; Prakash *et al.*, 2012; Rafalko *et al.*, 2012;

Selevsek *et al.*, 2011; Yocum *et al.*, 2010). PSA in blood may be increased due to benign conditions including prostatitis or benign prostate hyperplasia (BPH). However, modestly elevated PSA is blood at middle age is also strongly associated with metastasis or death from prostate cancer decades later (Lilja *et al.*, 2011; Vickers *et al.*, 2010). PSA-screening in clinical diagnostics can reduce cancer-related deaths but may also lead to over-diagnosis and overtreatment (Hugosson *et al.*, 2010; Schröder *et al.*, 2012). Hence, controversy remains regarding the merits of the PSA-test (Moyer, 2012; Vickers *et al.*, 2012), although it stays as mainstay in monitoring of therapeutic intervention, detection of disease recurrence or progression (Lilja *et al.*, 2008).

PSA is a prostatic kallikrein-related serine peptidase (KLK3) with restricted chymotrypsin-like specificity, which has a function in liquefaction of seminal coagulum by degrading the key gel-forming proteins SEMG1 and SEMG2 (Lilja, 1985; Lilja *et al.*, 1987, 1989). The enzymatically active form of PSA is a 237-amino acid single-chain glycoprotein with a molecular weight about 28 kDa (Christensson *et al.*, 1990; Lundwall and Lilja, 1987). The epithelial cells of prostate, regulated by the nuclear androgen receptor, can express and release of predominantly catalytic PSA at 5–50 μmol/l concentrations of into seminal fluids. In contrast, these levels in blood normally are normally a 1,000,000-fold lower (20 pmol/l), and PSA is catalytically inactivated in a covalent complex with α_1-antichymotrypsin (SERPINA3) (Balk *et al.*, 2003; Lilja *et al.*, 1991; Stenman *et al.*, 1991).

The high sequence specificity of transitions utilized in SRM assays together with the outstanding sensitivity of triple quadrupole MS instruments allow to apply this methodology not only for quantification but also to identify new isoforms, variants or even mutant proteins in biological samples. Existing genomic databases can form the base for selection of unique signature sequences of such proteoforms. Following theoretical (*in silico*) tryptic digestion and blast search in the most complete protein sequence databases, synthetic heavy-isotope labelled peptides can be obtained to be able to develop new SRM assays. These peptides are then spiked into selected biological samples and used as internal standards to identify and further on, to quantify novel forms of proteins.

As an example, a number of proteoforms of PSA were selected and SRM assays were developed following the procedure described above and a large number of clinical samples of seminal and blood plasma was screened using specific transitions. This first attempt to find novel proteins resulted in identification of a mutant PSA, which was never described at expression level before (Fig. 8.2) (Végvári *et al.*, 2013). Most importantly, this product of single nucleotide polymorphism (SNP) variant of PSA was discovered in 9 out of 72 clinical samples, carrying the non-synonymous mutation: L132I, which is located within the LSEPAE(L/I)TDAVK peptide. Due to the isobaric precursor and fragment ions, identical transitions were produced and observed in analysis of those specific samples. The peaks of LSEPAELTDAVK and LSEPAEITDAVK were baseline separated in the reversed-phase gradient used, clearly indicating that the LSEPAEITDAVK sequence is more hydrophilic having a shorter retention time. Since both of the two isoforms can be present in the same sample (heterozygous expression profile), the areas of both peaks have to be combined when quantifying the total amount of PSA. The population-based frequency of the allele A in exon 3 of the *KLK3* gene (dbSNP code: rs2003783) showed 10% worldwide, 8% in Asia and Europe, 14% in Africa and 11% in America as reported in 1000Genomes database (see Fig. 8.2). Similar frequency rate (12%) was observed in Swedish study cohorts used for re-sequencing and genotyping of all kallikrein genes (Klein *et al.*, 2010).

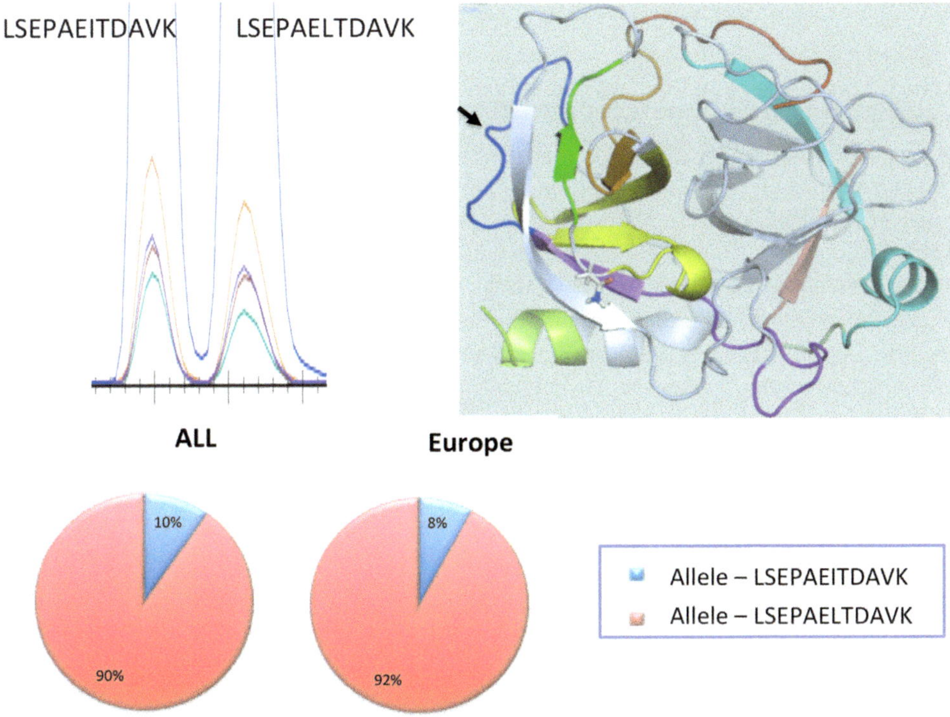

Figure 8.2 Identification of a novel mutant PSA in clinical samples by SRM assay. The SNP L132I has resulted in identical transitions of the peptides LSEPAELTDAVK and LSEPAEITDAVK, which were however chromatographically separated based on their hydrophobicity. The arrow indicates the loop in PSA with the point mutation. The allele frequency of this SNP product is about 10%.

The advance of nano-LC-MS/MS technology made it possible to process smaller sample volumes provided that the target proteins are present at low ng/ml levels, when analysis could meet high reproducibility. Many cases, however, sample enrichment is still required in order to reach LODs of important biomarkers in blood plasma samples. Advanced chromatographic systems can already provide high-resolution when combined with intelligent selection of fragments containing target molecules (Paik *et al.*, 2012) as well as immunoreaction enrichment of biomarkers of sub-ng/ml levels. Today's developments hold a great potential of becoming an optional platform for clinical analyses in the future (Anderson *et al.*, 2004).

Conclusions

The current clinical sample biorepository developments are of major relevance to the proteomics and genomics fields, as it outlines the standardization aspects of biobanking and the requirements that are needed to run future clinical studies that will benefit the patients where omics science will play a major role. A global view of the field is given where best practice and conventional acceptances are presented along with ongoing large-scale biobanking projects.

It is envisaged that the upcoming generation of personalized drugs for targeted and stratified patient treatment will break through in major disease areas such as lifestyle-related cancers, in particular cancers that have the highest mortality including a predisposing disorder and chronic inflammatory disease. In addition, cardiovascular diseases, neurodegenerative diseases such as multiple sclerosis, Parkinson's disease and Alzheimer's disease, obesity and diabetes are additional disease areas where a multitude of pipeline drugs are expected to make it as products into the market and the shelves of the local pharmacy readily available for patients. Mass spectrometric technologies can provide the 'phenotypic fingerprint' required for the concept of personalized medicine. Mass spectrometry-driven target biomarker diagnoses in combination with high-resolution computed tomography could provide a critical pathway initiative facilitated by a fully integrated electronic healthcare infrastructure system (e-Health).

Translating basic science to discover new and better treatments is an enormous challenge where the understanding of protein function(s), both as targets as well as biomarkers will bring great value in the near future to come. This includes the missing proteins (currently about 30% of the human proteome), which have not yet been identified on a protein level but predicted from DNA, or identified as transcripts. Another area where the C-HPP will have an important role to play is within the area of the future healthcare with drug and diagnostic developments for novel treatments of complex diseases. The future direction envisages that large studies will be accompanied by biological and molecular information with corresponding material from large numbers of patients and healthy persons, whereas the protein data from large biobank archives will play a mandatory role. Accordingly, there is a call for further develop value and treasure that biobanking holds with a global approval on the quality, standardization and the organization of modern units being developed for future use. There are also demands and expectations from national agencies on quality for a successful usage of biobank materials, particularly in the area of novel drug developed with accompanying diagnostics introduced into the hospitals.

List of abbreviations used

BBMRI, Biobanking and Biomolecular Resources Research Infrastructure; FDA, Food and Drug Administration; HUPO, Human Proteome Organization; HPP, Human Proteome Project; MRM, multiple reaction monitoring; MS, mass spectrometry; LC, liquid chromatography.

References

Addona, T.A., Abbatiello, S.E., Schilling, B., Skates, S.J., Mani, D.R., Bunk, D.M., Spiegelman, C.H., Zimmerman, L.J., Ham, A.-J.L., Keshishian, H., *et al.* (2009). Multi-site assessment of the precision and reproducibility of multiple reaction monitoring-based measurements of proteins in plasma. Nat. Biotech. 27, 633–641.

Anderson, N.G., Matheson, A., and Anderson, N.L. (2001). Back to the future: The human protein index (HPI) and the agenda for post-proteomic biology. Proteomics *1*, 3–12.

Anderson, N.L. (2005). The roles of multiple proteomic platforms in a pipeline for new diagnostics. Mol. Cell. Proteomics *4*, 1441–1444.

Anderson, N.L., and Anderson, N.G. (2002). The human plasma proteome – History, character, and diagnostic prospects. Mol. Cell. Proteomics *1*, 845–867.

Anderson, N.L., Anderson, N.G., Haines, L.R., Hardie, D.B., Olafson, R.W., and Pearson, T.W. (2004). Mass spectrometric quantitation of peptides and proteins using stable isotope standards and capture by anti-peptide antibodies (SISCAPA). J. Proteome Res. 3, 235–244.

Balk, S.P., Ko, Y.J., and Bubley, G.J. (2003). Biology of prostate-specific antigen. J. Clin. Oncol. *21*, 383–391.

Barnidge, D.R., Goodmanson, M.K., Klee, G.G., and Muddiman, D.C. (2004). Absolute quantification of the model biomarker prostate-specific antigen in serum by LC–MS/MS using protein cleavage and isotope dilution mass spectrometry. J. Proteome Res. *3*, 644–652.

Betsou, F., Lehmann, S., Ashton, G., Barnes, M., Benson, E.E., Coppola, D., DeSouza, Y., Eliason, J., Glazer, B., Guadagni, F., *et al.* (2010). Standard Preanalytical Coding for Biospecimens: Defining the Sample PREanalytical Code. Cancer Epidem. Biomar. *19*, 1004–1011.

Christensson, A., Laurell, C.B., and Lilja, H. (1990). Enzymatic activity of prostate-specific antigen and its reactions with extracellular serine proteinase inhibitors. Eur. J. Biochem. *194*, 755–763.

Domon, B., and Aebersold, R. (2006). Review – mass spectrometry and protein analysis. Science *312*, 212–217.

Enjalbert, Q., Girod, M., Simon, R., Jeudy, J., Chirot, F., Salvador, A., Antoine, R., Dugourd, P., and Lemoine, J. (2013). Improved detection specificity for plasma proteins by targeting cysteine-containing peptides with photo-SRM. Anal. Bioanal. Chem. *405*, 2321–2331.

Fitzgerald, J.B., Schoeberl, B., Nielsen, U.B., and Sorger, P.K. (2006). Systems biology and combination therapy in the quest for clinical efficacy. Nature Chem. Biol. *2*, 458–466.

Fortin, T., Salvador, A., Charrier, J.P., Lenz, C., Bettsworth, F., Lacoux, X., Choquet-Kastylevsky, G., and Lemoine, J. (2009a). Multiple Reaction Monitoring Cubed for Protein Quantification at the Low Nanogram/Milliliter Level in Nondepleted Human Serum. Anal. Chem. *81*, 9343–9352.

Fortin, T., Salvador, A., Charrier, J.P., Lenz, C., Lacoux, X., Morla, A., Choquet-Kastylevsky, G., and Lemoine, J. (2009b). Clinical Quantitation of Prostate-specific Antigen Biomarker in the Low Nanogram/Milliliter Range by Conventional Bore Liquid Chromatography-Tandem Mass Spectrometry (Multiple Reaction Monitoring) Coupling and Correlation with ELISA Tests. Mol. Cell. Proteomics *8*, 1006–1015.

Hancock, W., Omenn, G., LeGrain, P., and Paik, Y.-K. (2010). Proteomics, Human Proteome Project, and Chromosomes. J. Proteome Res. *10*, 210–210.

Hugosson, J., Carlsson, S., Aus, G., Bergdahl, S., Khatami, A., Lodding, P., Pihl, C.-G., Stranne, J., Holmberg, E., and Lilja, H. (2010). Mortality results from the Göteborg randomised population-based prostate-cancer screening trial. Lancet Oncol. *11*, 725–732.

Huttenhain, R., Malmstrom, J., Picotti, P., and Aebersold, R. (2009). Perspectives of targeted mass spectrometry for protein biomarker verification. Curr. Opin. Chem. Biol. *13*, 518–525.

Kato, H., Nishimura, T., Ikeda, N., Yamada, T., Kondo, T., Saijo, N., Nishio, K., Fujimoto, J., Nomura, M., Oda, Y., *et al.* (2011). Developments for a growing Japanese patient population: Facilitating new technologies for future health care. J. Proteomics *74*, 759–764.

Keshishian, H., Addona, T., Burgess, M., Kuhn, E., and Carr, S.A. (2007). Quantitative, Multiplexed Assays for Low Abundance Proteins in Plasma by Targeted Mass Spectrometry and Stable Isotope Dilution. Mol. Cell. Proteomics *6*, 2212–2229.

Klein, R.J., Halldén, C., Cronin, A.M., Ploner, A., Wiklund, F., Bjartell, A.S., Stattin, P., Xu, J., Scardino, P.T., Offit, K., *et al.* (2010). Blood Biomarker Levels to Aid Discovery of Cancer-Related Single-Nucleotide Polymorphisms: Kallikreins and Prostate Cancer. Cancer Prev. Res. *3*, 611–619.

Kosaka, T., Yatabe, Y., Endoh, H., Kuwano, H., Takahashi, T., and Mitsudomi, T. (2004). Mutations of the epidermal growth factor receptor gene in lung cancer: Biol. Clin. Imp. Cancer Res. *64*, 8919–8923.

Kudoh, S., Kato, H., Nishiwaki, Y., Fukuoka, M., Nakata, K., Ichinose, Y., Tsuboi, M., Yokota, S., Nakagawa, K., Suga, M., *et al.* (2008). Interstitial lung disease in Japanese patients with lung cancer – A cohort and nested case–control study. Am. J. Respir. Crit. Care Med. *177*, 1348–1357.

Kulasingam, V., Smith, C.R., Batruch, I., Buckler, A., Jeffery, D.A., and Diamandis, E.P. (2008). 'Product Ion Monitoring' Assay for Prostate-Specific Antigen in Serum Using a Linear Ion-Trap. J. Proteome Res. *7*, 640–647.

Landers, P. (2003). Cost of developing a new drug increases to about $1.7 billion. The Wall Street Journal, 8 December.

Lange, V., Picotti, P., Domon, B., and Aebersold, R. (2008). Selected reaction monitoring for quantitative proteomics: a tutorial. Mol. Sys. Biol. *4*, 222.

Lasso, R.A. (2010). The ethics of research biobanking. JAMA *304*, 908–910.

Li, Y., Tian, Y.A., Rezai, T., Prakash, A., Lopez, M.F., Chan, D.W., and Zhang, H. (2011). Simultaneous analysis of glycosylated and sialylated prostate-specific antigen revealing differential distribution of glycosylated prostate-specific antigen isoforms in prostate cancer tissues. Anal. Chem. *83*, 240–245.

Lichtenstein, P., De Faire, U., Floderus, B., Svartengren, M., Svedberg, P., and Pedersen, N.L. (2002). The Swedish Twin Registry: a unique resource for clinical, epidemiological and genetic studies. J. Inter. Med. *252*, 184–205.

Lilja, H. (1985). A kallikrein-like serine protease in prostatic fluid cleaves the predominant seminal vesicle protein. J. Clin. Invest. *76*, 1899–1903.

Lilja, H., Oldbring, J., Rannevik, G., and Laurell, C.B. (1987). Seminal vesicle-secreted proteins and their reactions during gelation and liquefaction of human semen. J. Clin. Invest. *80*, 281–285.

Lilja, H., Abrahamsson, P.A., and Lundwall, A. (1989). Semenogelin, the predominant protein in human semen. Primary structure and identification of closely related proteins in the male accessory sex glands and on the spermatozoa. J. Biol. Chem. *264*, 1894–1900.

Lilja, H., Christensson, A., Dahlén, U., Matikainen, M.T., Nilsson, O., Pettersson, K., and Lövgren, T. (1991). Prostate-specific antigen in serum occurs predominantly in complex with alpha 1-antichymotrypsin. Clin. Chem. *37*, 1618–1625.

Lilja, H., Ulmert, D., and Vickers, A.J. (2008). Prostate-specific antigen and prostate cancer: prediction, detection and monitoring. Nat. Rev. Cancer *8*, 268–278.

Lilja, H., Cronin, A.M., Dahlin, A., Manjer, J., Nilsson, P.M., Eastham, J.A., Bjartell, A.S., Scardino, P.T., Ulmert, D., and Vickers, A.J. (2011). Prediction of significant prostate cancer diagnosed 20 to 30 years later with a single measure of prostate-specific antigen at or before age 50. Cancer *117*, 1210–1219.

Liu, T., Hossain, M., Schepmoes, A.A., Fillmore, T.L., Sokoll, L.J., Kronewitter, S.R., Izmirlian, G., Shi, T., Qian, W.-J., Leach, R.J., *et al.* (2012). Analysis of serum total and free PSA using immunoaffinity depletion coupled to SRM: correlation with clinical immunoassay tests. J. Proteomics *75*, 4747–4757.

Lundwall, A., and Lilja, H. (1987). Molecular cloning of human prostate specific antigen cDNA. FEBS Lett. *214*, 317–322.

Lynch, T.J., Bell, D.W., Sordella, R., Gurubhagavatula, S., Okimoto, R.A., Brannigan, B.W., Harris, P.L., Haserlat, S.M., Supko, J.G., Haluska, F.G., *et al.* (2004). Activating mutations in the epidermal growth factor receptor underlying responsiveness of non-small-cell lung cancer to gefitinib. N. Engl. J. Med. *350*, 2129–2139.

Marko-Varga, G. (2013). Biobanking as the central tool for translational medicine. Clin. Trans. Med. *2*, 4.

Marko-Varga, G., Ogiwara, A., Nishimura, T., Kawamura, T., Fujii, K., Kawakami, T., Kyono, Y., Tu, H.K., Anyoji, H., Kanazawa, M., *et al.* (2007). Personalized medicine and proteomics: Lessons from non-small cell lung cancer. J. Proteome Res. *6*, 2925–2935.

Marko-Varga, G., Végvári, Á., Welinder, C., Lindberg, H., Rezeli, M., Edula, G., Svensson, K.J., Belting, M., Laurell, T., and Fehniger, T.E. (2012). Standardization and utilization of biobank resources in clinical protein science with examples of emerging applications. J. Proteome Res. *11*, 5124–5134.

Mok, T.S., Wu, Y.L., Thongprasert, S., Yang, C.H., Chu, D.T., Saijo, N., Sunpaweravong, P., Han, B.H., Margono, B., Ichinose, Y., *et al.* (2009). Gefitinib or Carboplatin-Paclitaxel in Pulmonary Adenocarcinoma. N. Engl. J. Med. *361*, 947–957.

Moyer, V.A. (2012). Screening for Prostate Cancer: U.S. Preventive Services Task Force Recommendation Statement. Ann. Intern. Med. *157*, 120–134.

Nishimura, T., Nomura, M., Tojo, H., Hamasaki, H., Fukuda, T., Fujii, K., Mikami, S., Bando, Y., and Kato, H. (2010). Proteomic analysis of laser-microdissected paraffin-embedded tissues: (2) MRM assay for stage-related proteins upon non-metastatic lung adenocarcinoma. J. Proteomics 73, 1100–1110.

Nyberg, F., Ogiwara, A., Harbron, C.G., Kawakami, T., Nagasaka, K., Takami, S., Wada, K., Tu, H.-K., Otsuji, M., Kyono, Y., *et al.* (2011). Proteomic Biomarkers for Acute Interstitial Lung Disease in Gefitinib-Treated Japanese Lung Cancer Patients. PLoS ONE *6*, e22062.

Paez, J.G., Janne, P.A., Lee, J.C., Tracy, S., Greulich, H., Gabriel, S., Herman, P., Kaye, F.J., Lindeman, N., Boggon, T.J., *et al.* (2004a). EGFR mutations in lung cancer: Correlation with clinical response to gefitinib therapy. Science *304*, 1497–1500.

Paik, Y.-K., and Hancock, W.S. (2012). Uniting ENCODE with genome-wide proteomics. Nat. Biotech. *30*, 1065–1067.

Paik, Y.K., Omenn, G.S., Uhlen, M., Hanash, S., Marko-Varga, G., Aebersold, R., Bairoch, A., Yamamoto, T., Legrain, P., Lee, H.J., *et al.* (2012). Standard guidelines for the chromosome-centric human proteome project. J. Proteome rRs. *11*, 2005–2013.

Park, A. (2009). Ten Ideas Changing the World Right Now: Biobanks. In TIME.

Picotti, P., Rinner, O., Stallmach, R., Dautel, F., Farrah, T., Domon, B., Wenschuh, H., and Aebersold, R. (2010). High-throughput generation of selected reaction-monitoring assays for proteins and proteomes. Nature Methods 7, 43–46.

Prakash, A., Rezai, T., Krastins, B., Sarracino, D., Athanas, M., Russo, P., Zhang, H., Tian, Y., Li, Y., Kulasingam, V., *et al.* (2012). Interlaboratory reproducibility of selective reaction monitoring assays using multiple upfront analyte enrichment strategies. J. Proteome Res. *11*, 3986–3995.

Rafalko, A., Dai, S., Hancock, W.S., Karger, B.L., and Hincapie, M. (2012). Development of a Chip/Chip/SRM platform using digital chip isoelectric focusing and LC-Chip mass spectrometry for enrichment and quantitation of low abundance protein biomarkers in human plasma. J. Proteome Res. *11*, 808–817.

Rodriguez, H., Tezak, Z., Mesri, M., Carr, S.A., Liebler, D.C., Fisher, S.J., Tempst, P., Hiltke, T., Kessler, L.G., Kinsinger, C.R., *et al.* (2010). Analytical validation of protein-based multiplex assays: a workshop report by the NCI-FDA interagency oncology task force on molecular diagnostics. Clin. Chem. *56*, 237–243.

Schröder, F.H., Hugosson, J., Roobol, M.J., Tammela, T.L.J., Ciatto, S., Nelen, V., Kwiatkowski, M., Lujan, M., Lilja, H., Zappa, M., *et al.* (2012). Prostate-Cancer Mortality at 11 Years of Follow-up. N. Engl. J. Med. *366*, 981–990.

Selevsek, N., Matondo, M., Carbayo, M.S., Aebersold, R., and Domon, B. (2011). Systematic quantification of peptides/proteins in urine using selected reaction monitoring. Proteomics *11*, 1135–1147.

Stenman, U.H., Leinonen, J., Alfthan, H., Rannikko, S., Tuhkanen, K., and Alfthan, O. (1991). A complex between prostate-specific antigen and alpha 1-antichymotrypsin is the major form of prostate-specific antigen in serum of patients with prostatic cancer: assay of the complex improves clinical sensitivity for cancer. Cancer Res. *51*, 222–226.

Surinova, S., Schiess, R., Huttenhain, R., Cerciello, F., Wollscheid, B., and Aebersold, R. (2011). On the development of plasma protein biomarkers. J. Proteome Res. *10*, 5–16.

Svartengren, M., Engstrom, G., Anderson, M., Hallberg, J., Edula, G., de Verdier, M.G., Dahlback, M., Lindberg, C.M., Forsman-Semb, K., Nihlen, U., *et al.* (2009). Twins studies as a model for studies on the interaction between smoking and genetic factors in the development of chronic bronchitis. Biochem. Soc. Trans. *37*, 814–818.

Végvári, Á., Rezeli, M., Häkkinen, J., Sihlbom, C., Carlsohn, E., Malm, J., Lilja, H., Laurell, T., and Marko-Varga, G. (2011). Bioinformatic strategies for unambiguous identification of prostate specific antigen in clinical samples. J. Proteomics *75*, 202–210.

Végvári, Á., Rezeli, M., Sihlbom, C., Häkkinen, J., Carlsohn, E., Malm, J., Lilja, H., Laurell, T., and Marko-Varga, G. (2012). Molecular microheterogeneity of prostate specific antigen in seminal fluid by mass spectrometry. Clin. Biochem. *45*, 331–338.

Végvári, Á., Sjödin, K., Rezeli, M., Malm, J., Lilja, H., Laurell, T., and Marko-Varga, G. (2013). Identification of a Novel Proteoform of Prostate Specific Antigen (SNP-L132I) in Clinical Samples by Multiple Reaction Monitoring. Mol. Cell. Proteomics: MCP *12*, 2761–2773.

Vickers, A.J., Cronin, A.M., Björk, T., Manjer, J., Nilsson, P.M., Dahlin, A., Bjartell, A., Scardino, P.T., Ulmert, D., and Lilja, H. (2010). Prostate specific antigen concentration at age 60 and death or metastasis from prostate cancer: case–control study. BMJ *341*, c4521.

Vickers, A.J., Roobol, M.J., and Lilja, H. (2012). Screening for prostate cancer: early detection or overdetection? Annu. Rev. Med. *2012*, 161–170.

Weston, A.D., and Hood, L. (2004). Systems biology, proteomics, and the future of health care: toward predictive, preventative, and personalized medicine. J. Proteome Res. *3*, 179–196.

Yocum, A.K., Khan, A.P., Zhao, R., and Chinnaiyan, A.M. (2010). Development of selected reaction monitoring-MS methodology to measure peptide biomarkers in prostate cancer. Proteomics *10*, 3506–3514.

Index

A

B

T

U